THE DROWNED PLACES

ALSO BY DAMIAN LE BAS

The Stopping Places

The Drowned Places

Diving in Search of Atlantis

DAMIAN LE BAS

Chatto & Windus
LONDON

1 3 5 7 9 10 8 6 4 2

Chatto & Windus, an imprint of Vintage, is part of the Penguin
Random House group of companies

Vintage, Penguin Random House UK, One Embassy Gardens,
8 Viaduct Gardens, London SW11 7BW

penguin.co.uk/vintage
global.penguinrandomhouse.com

Penguin
Random House
UK

First published in the UK by Chatto & Windus in 2025

Copyright © Damian Le Bas 2025

Damian Le Bas has asserted his right to be identified as the author of this
Work in accordance with the Copyright, Designs and Patents Act 1988

Every effort has been made to contact all copyright holders.
The publisher will be pleased to amend in future editions
any errors or omissions brought to their attention.

Penguin Random House values and supports copyright. Copyright fuels creativity, encourages diverse voices, promotes freedom of expression and supports a vibrant culture. Thank you for purchasing an authorised edition of this book and for respecting intellectual property laws by not reproducing, scanning or distributing any part of it by any means without permission. You are supporting authors and enabling Penguin Random House to continue to publish books for everyone. No part of this book may be used or reproduced in any manner for the purpose of training artificial intelligence technologies or systems. In accordance with Article 4(3) of the DSM Directive 2019/790, Penguin Random House expressly reserves this work from the text and data mining exception.

Typeset in 12/14.75pt Bembo Book MT Pro by Jouve (UK), Milton Keynes
Printed and bound in Great Britain by Clays Ltd, Elcograf S.p.A.

The authorised representative in the EEA is Penguin Random House Ireland,
Morrison Chambers, 32 Nassau Street, Dublin D02 YH68

A CIP catalogue record for this book is available from the British Library

HB ISBN 9781784743994
TPB ISBN 9781784744007

Penguin Random House is committed to a sustainable future
for our business, our readers and our planet. This book is made
from Forest Stewardship Council® certified paper.

MIX
Paper | Supporting
responsible forestry
FSC® C018179

The city and citizens, which you yesterday described
to us in fiction, we will now transfer to the world of reality.

Plato

Those who have once listened to the siren-songs
of the ocean bed never return to land.

Philippe Diolé

For Nan, who said
'The sea is true'

Contents

Map of Baia, Italy	xii
Map of Port Royal, Jamaica	xiii
Map of Pavlopetri, Greece	xiv
Timeline	xv
Of Rán and the Bells	1
The English Channel	
The Isle of Atlas	4
Plato's Atlantis	
Fathers	7
Shadwell Basin, London	
Stingray	11
Sussex Rock Pools, England	
Selkie-Skins	15
Sussex and Kent, England	
The First Breath	28
Orihuela Costa, Spain	
Atlantis Since Plato	41
Auschwitz; Byzantium; the 'New World'	
The Trace	53
Cala Cortina, Spain	

The Puzzle	65
Marburg, Germany	
Redemption	73
Wraysbury, England	
New Questers	86
The Straits of the Mediterranean; the Deep Sea	
Hermits	94
Widewater Lagoon, England	
The Lessons of Stone	106
Lake Van, Turkey	
My God! Seals!	115
Farne Islands, England	
Transfer to Reality	135
Hasankeyf, Turkey; Lion City, China	
The Floating Rocks	140
Santorini, Greece	
Palace of the Sea	152
Baia, Italy	
In the Beast	165
SS *Fortuna, English Channel*	
The Three Nails	178
Cabo de Palos, Spain	
The Ruins	188
Yonaguni, Japan	

11:43 a.m., 7 June 1692   194
Port Royal, Jamaica

The Turn of Poseidon   214
Achaea, Greece

He Walketh Through Dry Places   228
Pavlopetri, Greece

Future Atlantises   237
The Coasts of a Rising Ocean

Epilogue   244
Lokroi, Greece

Acknowledgements   249
Notes   251
Index   259

Sketch of Baia, Italy, from the author's journal

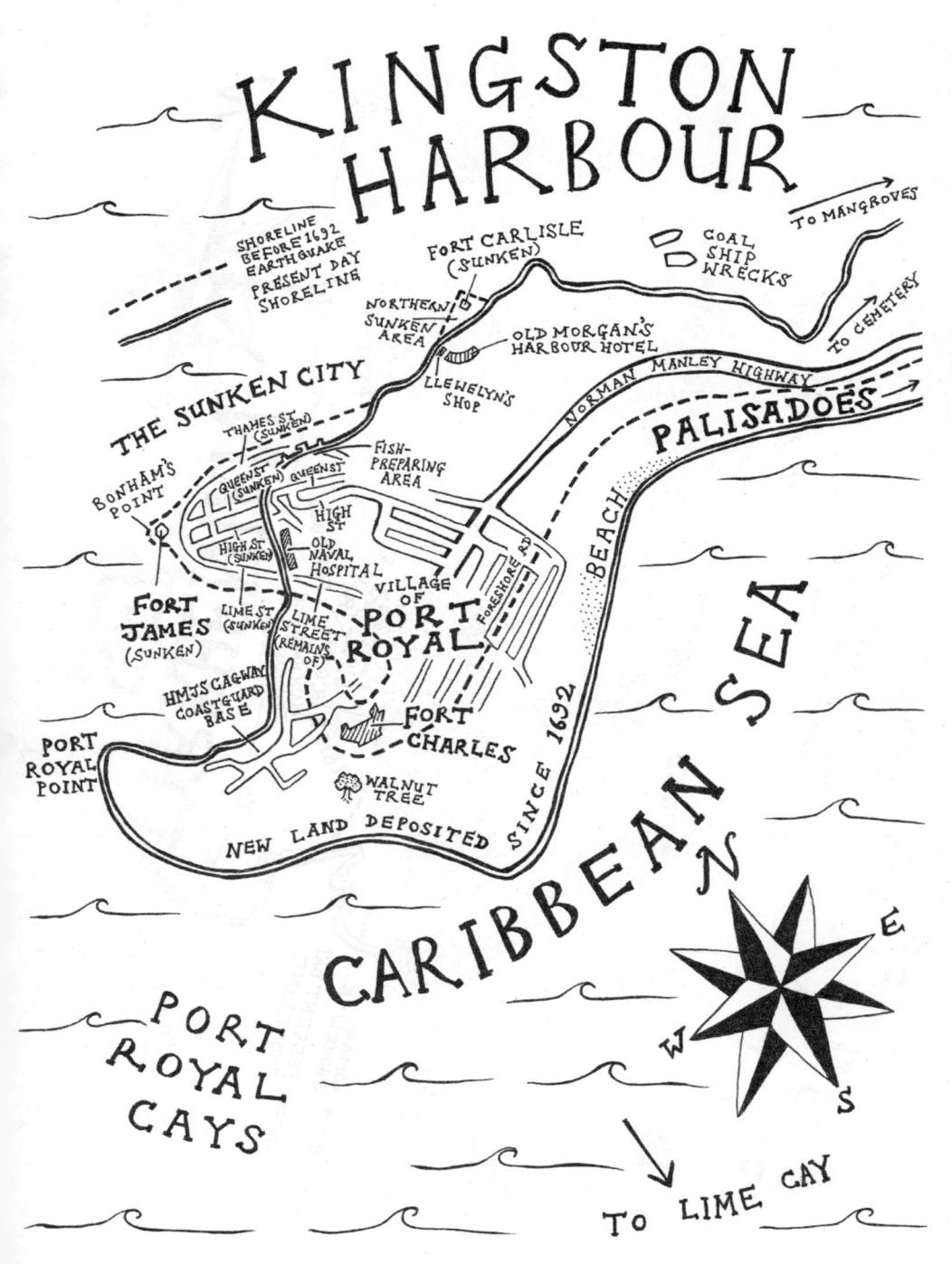

Sketch of Port Royal, Jamaica, from the author's journal

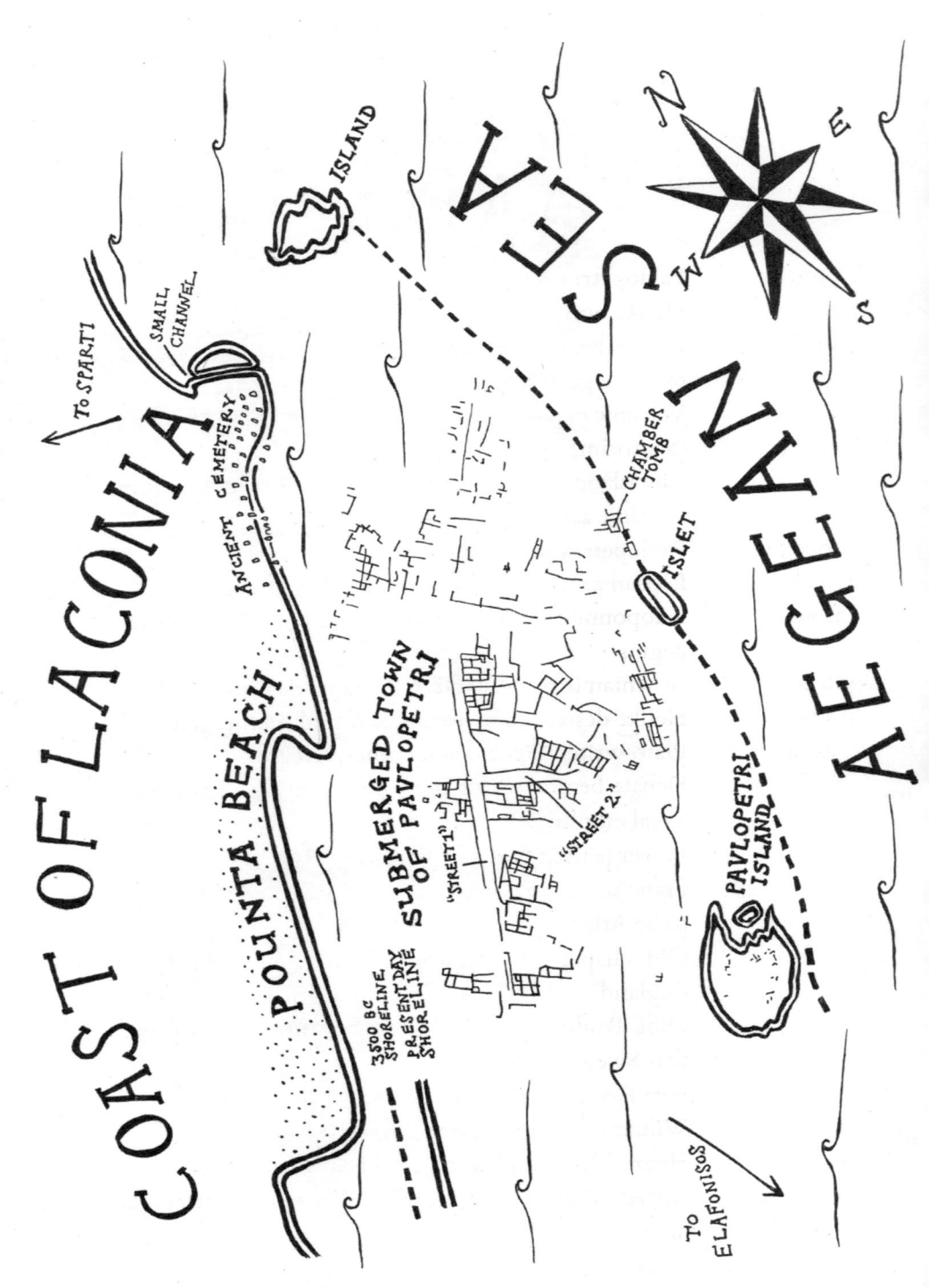

Sketch of Pavlopetri, Greece, from the author's journal

Timeline

c.3500 BC	Pavlopetri occupied
c.2650 BC	Helike occupied
c.1700 BC	Oldest known copies made of the Atrahasis epic, a tale of a great flood
c.1600 BC	Volcanic explosion kills all life on the island of Thera (Santorini), Greece
	The island's largest settlement, Akrotiri, is buried in volcanic ash
c.1000 BC	Pavlopetri submerged
480 BC	Persian army under Xerxes I invades Greece
431 BC	Peloponnesian War between Athens and Sparta begins
426 BC	Athenian fort on island of Atalanta destroyed
373 BC	Helike destroyed by earthquake and submerged
c.360 BC	Plato writes *Timaeus* and *Critias*, Greece
	Debate begins as to whether Atlantis may have been a real empire
c.800	Lower (eastern) part of Baia submerged, Italy
1552	Francisco López de Gómara declares the New World to be Atlantis
c.1600	Old chapel of Kingston lost to encroaching sea, England
1679–1702	Olof Rudbeck, in his *Atlantica*, attempts to prove that Sweden is Atlantis
1692	Port Royal submerged, Jamaica (7 June)
1793	William Blake publishes *America a Prophecy*
1878	Henry Fleuss invents rebreather system
1912	Alfred Wegener first proposes theory of continental drift
1916	SS *Fortuna* sinks, English Channel (22 October)

1931	Woods Hole Oceanographic Institution announces plan to find Atlantis
1943	Jacques Cousteau and Émile Gagnan invent automatic compressed-air diving lung that will come to be known as self-contained underwater breathing apparatus, or SCUBA
1953	*The Silent World* by Jacques Cousteau published
1959	Lion City flooded, China
1960s	Spyridon Marinatos and his team discover city of Akrotiri
	Robert Marx excavates ruins of Port Royal
1960s–1970s	'Doc' Edgerton and others carry out sonar surveys in search of Helike
1974	Marinatos dies at Akrotiri
1977	*Cayman Trader* sinks, Port Royal Cays, Jamaica (13 August)
1986	Yonaguni Monument discovered, Japan
1990–91	Largest known microbialites discovered in Lake Van, Turkey
1992	Pierre Vidal-Naquet publishes 'Atlantis and the Nations'
2000	Lost City hydrothermal vent field discovered atop Atlantis Massif
2002	Dora Katsonopoulou and her team discover Helike, Greece
	Graham Hancock links the Atlantis story to glacial melting events
2005	Robert Sarmast seeks Atlantis off the coast of Cyprus
2019	Hasankeyf flooded, Turkey

Of Rán and the Bells

The English Channel

Since my father's death, images of diving – and of committing things to the deep – had begun to stalk me. The paramedics who'd tried to resuscitate Dad had inserted a breathing tube into his mouth while they performed emergency decompressions on his chest. It was as if he'd drowned in his own bed. By the time I rushed into the room and found Mum kneeling distraught beside his body, the medics had already left, but the tube was still fixed in Dad's mouth, kept in place by unsightly rubber bands stretched round his head. It looked like a makeshift snorkel. In that first cold shock of grief, I had a vague thought that Dad was still trying to breathe through this straw like the divers of old, who drew air down into the water through hollow reeds. But there was no water. He was trapped beneath the surface of a different substance, submerged in airless death. There was also a dark hopefulness about the scene. I wondered why they would have left the tube in his mouth unless there was still a chance he might start to breathe again.

We buried him in his best clothes and his favourite piece of jewellery – a gold knot ring in the style once favoured by sailors for their weddings, a braid that would never come loose. When we lowered the black and gold casket into the earth, it felt like we were committing it to the protection of a deep place. I had always seen the interment of the dead as a final act, but now it felt like a gesture that demanded a response. I thought of people digging for chests full of riches or seeking them at the bottom of the sea. What was the sequel to this burial of treasure? Putting a breathing tube in my mouth and immersing myself in water felt like the closest I could get to Dad without following him to the grave. He would not have wanted me to. *Remember to respect the sea, son*, he always said. *Cos it ain't going to respect you.* But it was because of him that I needed to do it.

In centuries past there was a goddess of the sea, worshipped by people of the north. She was, in a sense, the sea itself: a personification of the deep, and a symbol of our need to revere it. She was thought to weave entanglements of enchanted nets and lines, ensnaring seafarers who had dared to sail or dive into the waves. Her name meant 'plunder', 'theft', 'seizure'. Perhaps in this name there was also a sense of revenge: of the sea striking back at mortals who invaded and despoiled her, whose weapons and ships polluted her world.

I was born beside the sea, and the ocean terrified and seduced me. I heard its voice in my thoughts. It called. It warned. It spoke to me in cyclic dreams of a world overrun by water, and in nightmares where I would find myself caught among the weeds and discarded ropes slung on the seabed. I gave the ocean's voice the name of the goddess. I called it Rán.

Submerged off the coast by my home town was an ancient chapel. On sunny weekends people walked the nearby beaches with their dogs, drank coffee and talked. I stared out to where the nubs of the ruins might be, masked by the grey-green water. Fishermen would sometimes swear they had heard the chapel's bells tolling, dull calls mooning up from the deep. No doubt it was the sound of inland belfries they could hear, but I imagined myself on one of those boats, bucking sharply at night, in the wild rain perhaps, without a stable horizon. What must it be like aboard a tiny craft with its fragile ribs, knowing that down on the seabed, the scoured terrain of the spider crabs and eels, a temple – if there was anything left of it – stood, underneath the dark slop of the waves? Would even the bells that called from the land not be terrifying, draped with a mariner's knowledge that one day they, like everything else, must be torn back into the ocean? That every coastal dwelling will one day suffer the fate of Atlantis?

What was Atlantis? It was the world I dreamt of under the sea, strange even to Rán, because it was one that human hands had left behind. It was a place of ruins: columns, arches, towers, walls caressed by soft blue light, invaded by fishes, wide-eyed at the broken palaces where waterweeds grow, where crabs instead of spiders wait in the crevices and a shark swoops through a door where people once walked. The drowning of a civilisation cannot help but set the stage for another life. Perhaps you have dreamt of going there, where our kind no longer belongs. It would be not only a trespass, but a return.

That was my dream: to go to the drowned places. It was the counterweight to my fear of Rán, imagining that the beauty might be worth it, teaching me things that I could not learn on the land. It meant escape and freedom, discovering inner strength and a new understanding of our world; which is, in truth, a sea-world, where the land is a blip in the blue. I followed the call of the bells to that place, and to the many meaning of Atlantis, as I hope you might. It led to the strangest of healings.

The Isle of Atlas

Plato's Atlantis

Long before our time, before even the time of the people who spoke of Rán, there was a legendary place beyond the sea.

This was some 11,500 years ago, in an age before even the oldest of the great empires of antiquity were born: when Babylon, Egypt and Rome were not yet whispers on the lips of gods. A vast island – a continent, perhaps, by our reckoning – the size of western Asia and North Africa combined, lay in the Atlantic Ocean, beyond the Pillars of Hercules[1] that separated it from the Mediterranean Sea. From north to south it stretched from the latitude of Cadiz far down the west side of the African continent. Springs and fountains danced across its wide and fertile plains, rich harvests grew without effort, and a young woman who had been orphaned on the cusp of her adulthood lived on a hill at the island's centre. Her name was Cleito.

The sea-god Poseidon had fallen in love with Cleito, and to keep anyone else from touching her, he raised concentric rings of hills and moats around her home. Before the invention of ships, no mortal men could pass Poseidon's defences, so he alone fathered Cleito's children. The eldest was named Atlas: not Atlas the Titan, whose shoulders supported the heavens, but Atlas the demigod, first king of the island which from then on bore his name: *Atlantis nesos*, the Isle of Atlantis.

Atlantis was blessed with many qualities which enabled its people to swiftly rise to power. Its rich lands meant they never hungered; its great size meant they never wanted for room; Poseidon's moats made it formidable in the face of attack. With natural resources of stone and precious metals, the Atlanteans built a temple, a canal and a harbour; and with a great navy they subjugated all whom they met along the coasts of the Middle Sea. This Atlantean domination was the first seaborne empire the world had ever seen.

But far to the east of Atlantis, on a peninsula that lay just outside its sphere of conquest, another power was rising. Just behind a notch of coast in the land we now call Greece, a frugal tribe of farmers and soldiers had founded a little city-state called Athens: an inland realm, without ships or men of the sea. This was a 'primeval Athens', the ancestor city of the Athens of the golden age that would arrive millennia later with its philosophers, playwrights, statesmen, sculptors and poets. This primeval Athens was a simpler place. It did not reach overseas to spoil and predate on the lives of others. Clear of the nautical reach of Atlantis, with enough food to feed itself, and enough soldiers for its defence, the city lived by its own laws. It had no colonial ambitions. It had no hubris. It was strong, austere and fair.

It seemed fated that Atlantis and Athens would eventually collide, and they did. The war climaxed in a terrible battle with enormous casualties. Athens prevailed; but, in the aftermath, there was a great earthquake. The land was riven in pieces, and the armies of both sides were swallowed up into the earth. In a single night and day, the mighty island of Atlantis sank, together with all its civilians and survivors, its sciences and its armaments, its temples and its animals, into the sea.

All that was left was a treacherous area of mud and ooze lying just beneath the surface of the water, notorious to sailors as a death-trap. Atlantis would be memorialised in the form of the ocean that, to this day, bears its name: the Atlantic, stretching from the Pillars of Hercules to the New World.

As for its history, it fell entirely out of memory, as did the exploits of its valiant foe, primeval Athens. Except, that is, in one place – the city of Sais in the Nile Delta of Egypt, among the wise class of priests whose depth of lore became the envy of the younger Greek world. In Egypt the story of Atlantis was handed down for some ninety centuries, until it was finally conveyed from one of these priests, via several Greek go-betweens, to Plato, one of the greatest philosophers of classical Greece.

Atlantis is no more. It lives on only as a story, a cautionary tale of empire and what happens when power is overreached by pride. There are some who believe it was the seed of later civilisations, that the far-flung peoples of the world must surely share some old and common root in the great lost island. The sea has taken Atlantis, but Atlantis is the name that will not drown.

This, at least, is what some would like to believe.

Fathers

Shadwell Basin, London

Time spreads like dye in cold water. Sorrows drown fast. Each drawing apart of the hands was like opening thick green curtains into a different way of being: prying open another realm that constantly recedes from reach.

Shadwell Basin is an old dock on the north bank of the River Thames, nowadays closed off to larger boats and surrounded by red and grey blocks of flats with steeply angled roofs. Its waters are thick with algae. On its bottom lurk shopping trolleys and motorcycles, now the furniture of fish: moustachioed Eurasian carp and scoop-mouthed pike. Online, men share pictures of the animals they have pulled out of the Basin. The same individual fish, recognisable by their missing scales and their peculiar tints of brass and gold, are, like petty criminals, repeatedly hooked and tackled and photographed, then thrown back into their shady underworld.

Swimming in the Basin is against the local laws, and there are signs that warn of its forty-foot depth, of the unknown hazards on the bottom, and the risk of catching Weil's disease from the urine of local rats. Weil's can cause haemorrhaging in the lungs, and if you catch it the chance of death is maybe 5 to 10 per cent. But the major cause of human demise in the Basin has been, predictably, drowning. Since

2010 there have been at least three fatalities, adding a haunted layer to the green water's ghoulish look. In spite of all this, the Basin calls to the swimmer. Its quivering surface holds a promise of cool and current-free water, the perfect place to revive a skin that has been kept too long in the dark, and spark a bit of rhythm into winter-addled muscles.

It was a sharp, clear morning. I mooched along the gangway of planks to the path that ran around the dock's edge and got changed into my trunks. Too scared to dive in head-first, I clambered down the slimy ladder and let out a silent gasp as the wintry water closed up to my neck, then swam breaststroke for a couple of minutes. Into these minutes was crammed more vital sensation than a body might feel in a fortnight spent indoors. And as time slows down at the thrill of this dazzling psychic and bodily stimulation, it is hard to assess how long you have been swimming for. In cold water the seconds disperse like ink, and the grime of quotidian worries dissolves away. Something in you is desperate to climb out and escape the cold, but each new stroke through the water brings you closer to the promise of a mysterious reward. Delving into another realm that constantly recedes from reach but which – with each attempt to touch it – stills another unwelcome thought, dissolves a little more of the anxiety that encrusts the modern mind. It is almost absurd that an experience so transformative does not bring about more visible change in the human body. When I got out I was shocked to find that I hadn't turned green, but only a little flushed, the red of a mullet. After towelling off, I dressed and hurried back along the boardwalk. I had an appointment to meet my old friend Jamie Berger.

I used to live with Jamie in a London house full of gibbeted mice, Victorian flagstones and the smell of three-day parties. He had taught Sino-Japanese history at Harvard University, launched a restaurant known for smoked meats and terse bourbon cocktails, and had a sword, a petrified megalodon tooth, and an ancestor whose portrait was on the Confederate hundred-dollar bill. Our house lay a mile to the south of the Thames. It sat in a near-perpetual gripe of cold: sometimes the water in the toilet bowls froze solid. In those days Jamie's breakfast was always a foam-topped quadruple espresso and a pair of hand-rolled cigarettes, and my breakfast was the same as Jamie's. Ten years had passed. We'd both stopped smoking now and shared a bond forged partly from the bitterness of the grudgingly reformed.

We met at the Grapes pub on Limehouse Reach, a beerhouse since the reign of Queen Elizabeth I. The bar was dark. Entering felt like an intrusion, and the yellow gloom made innocent middle-aged Londoners look like smugglers doomed to hang. The staircase to the bathroom was almost as steep as a crow's-nest ladder, and everywhere there was the perfume of brackish moisture: of water that smelt undrinkable but which contained the promise of motion, adventure, danger, ships and risk. From the back of the pub jutted a balcony overlooking the spill of the Thames, and on the riverbank below a slender gallows could be seen: its root was black with bladderwrack weed, clutched by the river's dark life. In *Our Mutual Friend*, Charles Dickens described 'a tavern of dropsical appearance . . . the whole house impended over the water but seemed to have got into the condition of a faint-hearted diver, who has paused so long on the brink that he will never go in at all.' He is thought to have had the Grapes in mind as he wrote.

Jamie's father and mine had died not long apart. There was something of sons adrift about our meeting. Jamie ordered two pints of amber ale and we said 'cheers'. Through the balcony doors of the pub the sounds of the Thames, its slaps and plips against the river wall and, behind and beyond them, its mighty hush, washed in.

'What did your father die of?' asked Jamie.

'A stroke,' I said. 'Or as it says on the post-mortem paper, a spontaneous subarachnoid haemorrhage.'

Our glum looks were lightened slightly by the absurd-sounding medical language. To a doctor these words explain what happened. To most people, they conceal it. That can be useful when breaking bad news: when something has happened that isn't nice to describe in simpler words. A blood vessel in Dad's brain had burst and caused a lethal flooding of the *pia mater*, the 'tender mother', a delicate layer of brain tissue that cannot survive immersion. Thinking about this next to the sound of the water, it occurred to me that it had been a kind of drowning after all.

I knew Jamie was a diver. I told him that since Dad died I had become fixated with the idea of diving, and taken to poring over pictures of underwater ruins.

Jamie was quiet for a moment. 'Have you ever dived?' he asked.

'No. Is it difficult?'

'Can be.'

'But you enjoyed it?'

'Oh, enormously!'

'What in particular?'

Jamie paused and smiled as memories and feelings came back to him. 'There's such a tremendous sense of transgression. Of truly having entered another world. The weightlessness people talk about, that definitely happens, and it's remarkable. Though it's not quite weightlessness exactly, but something else . . . more like being fully supported by the water. But the magical thing is the sense that you have somehow gone somewhere you were not . . . not designed to be. A feeling of having cheated nature. And the animals know you're this strange thing and they investigate and they wonder what you are. There is really nothing like it.'

I sat back on my stool, astonished.

'And fingers crossed, nothing goes wrong!' he said.

Stingray

Sussex Rock Pools, England

In the pools, green seaweed lay clotted and draped over chunks of rock in a way reminiscent of old ruins, where mosses and lichens encamp on tumbled blocks of stone.

Since I first learnt to walk I'd been observing the tides of the southern English coast, puzzled at this change that came over the shore and which seemed to have no equivalent on land: a contrast like night and day, but made by liquid instead of light. I waited for low tide and watched reruns of *Stingray*, an old children's programme in which puppet aquanauts voyage through technicolour seas in a blue-finned whale-esque submarine, encountering silver-skinned fish-people who lived on the ocean bed. Once the sea had retreated – it went out incredibly far at spring tides, for reasons that were a mystery to me – I would spend hours searching about in rock pools, mustering the strength to lift heavy, barnacled stones and, when I felt brave, squeamishly brushing aside the gutweed, thongweed, bladderwrack and short kelp, slimy life forms with unglamorous names. Too young and impatient to notice the shy, armoured motion of limpets and winkles, too finicky to fall for the heart-shaped air chambers of the olive-drab spiral wrack, at first I was looking for anything that had obvious senses, and that moved at a speed comparable to my own: shiny green shore crabs; spiny orange spider crabs; brown squat lobsters; shocked-looking, mudskipper-like blenny fish; and see-through prawns – organs on show through glass flesh – that darted backwards from my touch as if terrified.

There was something that kept bringing me back to the rock pools. They didn't fully belong to the sea or to the land, but there was more to it than that. They never quite seemed to look 'natural'. They looked like the site of a forgotten defeat, of reinhabited devastation: a smashed place, the wreckage of something that had once been taller and grander, and which now housed an entirely new kind of life and purpose. Marine algae curdled in pools, and flat fronds of seaweed lay slung over boulders like vines on abandoned temples. Was this nothing more than a ruin of geological chaos? Sometimes I thought I saw lines of stones, and wondered if they might be the remains of buildings. Surely not: but occasionally the wooden rows that had once been groynes could be seen, worn down to stumps like sperm whales' teeth, poking up out of the sand as though from a shallow grave.

As I got older these boyish daydreams found solidity in a single word: Atlantis. Atlantis meant underwater ruins, temples with white stone columns. It meant a bearded sea-king, a merman perhaps, with a trident in his hand, ancient powers and treasure. But not everything the word conjured was hoary with age. Atlantis could also be high-tech, futuristic: submarines and divers exploring the secrets of the deep. A sci-fi civilisation under the waves. I had no idea why this word evoked such a bizarre range of images, some of which were at odds with each other. The point of connection seemed to be the ocean. Without the sea, the columns were just the remnants of classical Greece or Rome. Without the sea, the man with a trident was just a statue; the marvellous city might have been Fritz Lang's Metropolis. It was the sea that held it all together: it was the world in which these ideas had to exist in order to qualify as Atlantean.

This Atlantis had no creator. It never occurred to me that someone might have consciously invented the word, and the thought that an ancient philosopher could have coined it as part of a literary allegory was beyond me. I saw Atlantis everywhere. It was in cartoons and spelled out on the pull-out awnings above cafés. Surely the word must have been around from the dawn of civilisation, or at least since the days of classical Greece. Restaurants called ATLANTIS flew the blue and white stripes of the modern Greek flag, and had white plastic pilasters with Ionic capitals beside the front door. There was something very Greek about Atlantis, but that was of far less interest to me than its something-of-the-sea.

My childhood beachcombing and paddling made me yearn for greater depth. As I got older I became a braver swimmer, wading out at first, then splashing in the calm shallows during the summer, eventually diving headlong into cold winter waves. I was addicted to the icy thrill that gave way to a clear-minded, wordless calm, and finally to a sense of semi-numbed animal presence that my friend Simon Bracken – a year-round outdoor swimmer – refers to as 'your second body'. Like others, I had become obsessed with 'wild swimming', which felt like a plunge into raw reality, unlike the bleached and tepid water-cube of the public indoor pool. I was hooked on the sea, and bereft when away from it for long. I missed how it put my troubles in perspective. Swimming provides the only means for humans to travel with our eyes just above surface level, our gaze brought low, almost level with a pond skater's. Combined with the way it makes us peer out over a constantly shifting terrain, a reality melting and recasting itself before we can catch its shape, we find that swimming humbles us amid the flash of the slosh and sun. No treeline, rarely any property to be seen: a flag of swimmers would be two simple swathes of colour, the blue on the green, with a restless fault line in between. It's impossible to paint, but it can be borne in mind.

But even in the sea, I was tethered: always reined back to the shore, tied to the surface and to the land by the requirements of my lungs and skin. And when I finally mustered the courage to dive down to the bottom, forcing my eyes to stay open in the pale brown-yellowish murk of the swirling water, I could make out little and struggled to maintain a gaze through the stinging brine. Nature programmes with underwater photography became unbearably taunting. I remembered my earliest fantasies and dreams. In one, my primary school had flooded from floor to ceiling, but somehow I was able to swim its corridors without surfacing for a breath. In another, a merman with a finned green head and huge fish eyes strode up the beach of my home town. He was wearing a mask attached by tubes to a pressurised tank of seawater on his back, allowing him to breathe in our toxic world. He was a fish-man 'diving' on land; a sort of anti-scuba diver. *He's come from Atlantis,* my childhood-self imagined, *and I'll go there one day.*

Selkie-Skins

Sussex and Kent, England

'They'd make you take all weights and your gear off and chuck it in the pool. Then they'd say, "Now go down and get it." And you had to dive down and bring it all back up. Easier said than done.'

A few weeks after my dip in Shadwell Basin and chat with Jamie, a cold March morning came round. The crows were hunched on the beach huts. I had set out for the local second-hand emporia, seeking cheap bookcases to impose some order on the mountains of records, books and CDs that Dad had left behind. The search took me to deserted industrial estates where great barns of corrugated grey metal housed every type of cast-off appliance except the ones I needed. The staff were helpful, but only when it came to warning me that mine was a pointless quest. Some of them teased me.

'Everyone's after shelves, mate. Funny, we just sold a load about, ooh, half an hour ago. Never mind!'

I minded.

The last place on my list was the Salvation Army furniture shop, a single enormous strip-lit room full of floor lamps and tables and chairs, run by a small team of Christians. Clouds were blowing in off the sea and I did not expect to find what I wanted. But they had a good bookcase. It was old, wooden, sturdy and cheap. I paid for it, carried it out with one of the volunteers, slid it into the back of the van and slammed the doors.

I turned around. My eyes snagged on the window of a fishing-tackle outlet. There was a mannequin standing in it, holding a fishing rod and wearing nothing but a pair of polarised sunglasses. That's when I noticed the next business along was a dive shop.

On the hoarding above the door, SCUBA CENTRE was written in huge dark letters. There was a flag in the window that said PADI, the acronym for the Professional Association of Diving Instructors. It bore a logo, a red caricature of a scuba diver holding a flame and swimming around a blue globe. It reminded me of the opening of Jacques Cousteau's film *The Silent World*, in which a group of divers, carrying blazing flares in their hands, swim down into deep black water. I had never forgotten that scene, those images. They represented a potent blend of the futuristic and the mythical. With their masks and air cylinders, the divers looked a bit like space travellers, but not quite. The long fins on their feet and the bubbles frothing away from their heads made it clear they were underwater. The furious torches they held, miraculously burning in water as their bearers forged ahead in the darkness, made them reminiscent of Greek gods journeying down to Tartarus, the Titan prison under the sea. The window display was full of diving-related items: posters advertising various courses, and a plastic torso dressed in brand-new diving gear. In one window, blazoned on to the glass in blue vinyl letters, it said LEARN TO SCUBA-DIVE.

For years I had thought about scuba diving and never acted on it. But now it seemed like fate.

You'd have to be off your head to do that, I heard Dad's voice say in my head.

Maybe, I thought, then slowly walked up to the door.

As I raised a hand to open it I felt a chill of excitement and fear, a fear that wouldn't have made sense if this were a shop dedicated to the pursuit of a 'normal' sport. Philippe Diolé, in the golden early years of the popularisation of diving, had described it as 'at the same time a sport, an instrument of scientific research and the means towards a "new human condition"'. For me it was also a lifelong dream and a cultural taboo.

Among Romany Gypsies the power of water is feared. There are likely several reasons for this. My family were inland people, travellers of fields and roads, only recently arrived at the coast. For them the dramas of the sea were a wild and dangerous vista, their ancestors' maritime voyage to Britain a memory long since forgotten. And there was more to it than that: the seeds of Romany culture come from the Indian subcontinent, and in my relatives' terror there was perhaps an echo of the Hindu reverence for the depths of the ocean and its resident monsters, the *houglis*. Now I stood before a portal to the sea, advertising itself to novices: people who, whatever their background, had yearnings similar to mine. The door represented an entry not into a leisure outlet, but into the ideas of risk and learning and the mysteries of the planet.

My fear did not subside. I opened the door and raised my eyes.

I had never seen such a thrilling and alien mix of things to buy. Everywhere there were garments and devices made out of futuristic valves and rubber tubes and shining steel. The place felt like a cross between a bicycle shop, a sportswear store and a specialist car-parts emporium, and yet it was none of these: it was an armoury of a unique way of life, an explorer's stockpile of the sea. There was nobody behind the counter, but I could hear a voice on the phone as I slowly walked around.

Among the few things I recognised were the fins, which I would have called flippers if someone had asked. Everything else was new to me. On the left-hand side was a rack of what looked like police stab vests, black, with flashes of colour. Underneath these strange jackets were masks and snorkels, some clear and edged with pink and turquoise, some black and grim with a single glass window, like something a spear-fisherman would have worn in the 1950s. There seemed to be a division implicit in the kind of mask someone would choose: did they see this as a fun hobby, or as something more serious? I gravitated towards the black masks, dark countenances glowering from their shelf. *Imagine staring through one of those at an underwater ruin, or a shipwreck full of eels*, I thought. The equipment seemed to feed these imaginings of adventure.

In the centre of the shop there was a vitrine full of small implements: underwater cameras, waterproof torches, short knives with plastic handles, and others with squared-off tips and notches in their blades – not weapons but tools. I was intrigued: there was a beauty to their utility, built for situations of which I had no understanding. I wanted them.

A wall to the right of the counter was covered in footwear: neoprene boots with thick rubber soles and treads like vehicle tyres. The wetsuits were familiar but behind them were racks of a very different kind of suit. They were bulky, cumbersome, and hung on specialised chunky hangers built to take their weight. Some had plastic rings around the necks and wrists. The legs terminated in socks with rubber soles, and on each one's chest was a round black thing like the petrol cap of a car. These suits didn't look designed for the water like the wetsuits. They were the sort of thing a firefighter or a Formula One driver or even an astronaut would wear. I could picture first responders grabbing these things in a terrible emergency. Where the wetsuits were svelte and figure-hugging, like the uniforms of movie superheroes or Olympic acrobats, these were the opposite. The lines of the human form would be lost in them, an irrelevance behind a wall of technical fabric built to shield its wearer from disaster.

The idea that people needed such formidable armour to go underwater was frightening. Maybe I had been right to be so scared of it. And why was everything black? Wouldn't it be a better idea to wear bright colours in the sea?

I walked back to the counter, which was glass-topped and full of treasures: logbooks, luminous wristwatches, weird little digital devices. Behind the counter was a door which led through to a large back room with a concrete floor. It was full of more suits – endless suits – and dozens of 'oxygen tanks' as I would have called them, painted white with black and yellow markings: a label of danger, as with nuclear weapons or poisonous fish. They looked heavy; it was laughable to think of someone walking or swimming about with one of these things attached to them.

My gaze went up to a wooden plaque above the door. It said MASTER SCUBA DIVER ROLL OF HONOUR, and was adorned with the names of those who must have attained this rank.

I didn't know how high up that made them in terms of the wide world of scuba diving, but I was immediately jealous. So that was the reward for strapping a big explosive canister onto your back. *Man*, I thought. *Imagine having your name on that. Imagine being able to call yourself a 'master scuba diver'.* I noticed that my mouth was slightly open.

'Can I help you?'

The phone call had ceased and someone was now standing behind the glass counter. She had a friendly, lightly tanned face and possessed the unmistakable glow of a person fulfilled by adventures. She waited for my reply, but with the contained energy of a busy and competent worker: someone who only ever stood still in the gaps between important tasks.

'Er, yes . . . I've been thinking about learning to dive. Maybe?'

'Great! Well, you've come to the right place, then, haven't you?'

'Um, yes!'

'Ever done a try dive?'

'No . . .'

'Any snorkelling?'

'A bit.'

'Good.'

These three questions qualified me for the sales pitch. She told me how much it would cost to do the basic open-water diver course, and the dates of the next one. It would be two days' study and practice here, split between the classroom and the pool, and two days' assessment in open water in a lake at Wraysbury. Wraysbury was not far away and yet, like the shop, I had never heard of it. This produced in me the feeling of a semi-hidden world being slowly revealed, simply because I had summoned the courage to open a door and ask.

'Have a look over this.' She handed me a large document packed with small text and a very long list of questions. Each one had YES or NO options, with accompanying boxes to tick. 'You should be answering "no" to all of these. It's just checks we have to do.'

I started to study the questions intently. The form seemed to ask about every medical affliction under the sun.

I couldn't believe anybody alive could have such perfect health as to be able to honestly answer 'no' to all these questions.

'Are you concerned about any of these?'

'Um . . . No, not any one in particular. There's just so many things!'

'Right, this form is just to say that you are healthy enough to dive. Ultimately it is an extreme sport.'

Extreme, I repeated internally.

'As in, even though it is safe and fun there will always be a certain amount of risk. So unless any of those particular questions relates to you, you don't need to worry.'

Her explanation made me realise something. I wasn't concerned about any of the questions on the form. It was the form itself. It was the reality of ticking all these boxes and signing this waiver. That would make it all become real. To sign this would be to stop dreaming and wondering, and to start down a path that meant spending time underwater.

I was floundering. She noticed.

'I mean, if you're having any reservations, you should talk this through with your doctor. We'd recommend that. For anyone who's wondering about the dangers.'

'I think maybe I'd like to do that actually, yes.'

'That's fine. In that case, do you want to take that away, speak to your GP and have a think? We run courses regularly, but if you want to get on the next one it's happening soon. So probably best not to leave it too long. There's limited spaces.'

'I think I'll do that, if that's OK.'

'Of course it is! No problem.'

My situation now resolved, she returned to the back room with its suits and scuba tanks. I turned around and walked back out of the shop.

Out in the street the cold March wind blew strong. Everything seemed grey. A few hundred yards to the south the sea growled and rumbled. I could smell it, a new world I was too afraid to visit.

I realised that what I had perceived as pressure to act quickly had simply been due to the fact that they do this all the time. People learn scuba diving every single day, and for those who teach them, this fact is not momentous. They get on with it. They ask you the necessary questions and sign you up. I had no health worries that precluded me from completing that form and enrolling on the course. So what had stopped me?

It was obvious. My shoulders sank at the realisation. I was afraid.

As I drove home, the bookcase shifted from side to side on the dusty plywood floor of the van. It made a dry sound, a sound of boredom and of days that passed by in arid tasks. It was a sound of motion towards domesticity, of doing the least interesting thing you had set out to do, and of never having the courage to take a path at the end of which were things you could not know. It was the least valiant sound I had ever heard.

I got home and sank down heavily into the worn side of the couch where my dad used to sit. I closed my eyes and breathed out. I had dreamt of finding Atlantis, and I couldn't even sign the form that would start me off on the journey. I was crestfallen.

Perhaps that is the punishment for cowardice. You must stay in your current state. You must stay *here*.

*

In the days that followed my exit from the dive shop, I felt deflated and depressed. My dreams had hit the rocks of fear and I couldn't dislodge them alone. If I was going to get out of this doldrum, then I needed to talk to someone who understood; another father. I had to talk to Gord.

My wife Candis's stepfather, Gordon, had been a recreational diver for many years. I knew of his diving past, but he wore it lightly, only showing it by way of the odd short and grizzled remark when the topic of scuba came up on the TV. He had stood mostly in silence watching news footage of the young Thai football team being rescued from a cave complex in Chiang Rai. At the end of a long and dramatic segment about the team of Thai and British divers who undertook the operation, one of whom – 37-year-old former Thai Navy SEAL Saman Gunan – died during the rescue, Gord uttered just one word: *Brave*.

On the mantelpiece was the only photo I'd seen of Gord together with his parents and his late brother Jack, a charismatic veteran diver who was at the centre of many family tales of the deep. With his broad chest, wide smile and fondness for diving blueholes – marine caverns open to the surface that look like dark portals to other worlds – Jack became a figure of inspiration for me, an aquanaut legend who probably would have been bemused to think anyone saw him like that. Jack had died a year previously and the photo seemed immovable now. It showed them all arm in arm having just disembarked from a speedboat beached on the Cornish coast. They had been diving. They wore wetsuits and unforced grins: there was a glow of pride and achievement. Apart from the neoprene and the waves lapping at their feet, it could have been taken at a baptism, a graduation or a wedding.

Gord had learnt to dive in the 1970s and 80s with the British Sub-Aqua Club. He did his training mostly in lakes and the fickle sea off the coast of Kent, England's most south-eastern county, which juts out into the ocean where the North Sea meets the Channel. His recollections were intimidating. Many of his instructors were hardened divers whose attitudes had been shaped by military experience: to them diving was not a sport, but a discipline, and a dangerous one which was not to be undertaken lightly. If they were going to certify someone as competent, then first they were going to put them to the test.

Things had started off in a swimming pool, extra deep for dive training: thirteen feet, double the usual six and a half at the 'deep end' of a public pool.

'They'd make you take all weights and your gear off and chuck it in,' said Gord. 'Then they'd say, *Now go down and get it*. And you had to dive down and bring it all back up. Easier said than done.'

I wasn't sure I'd be able to do even this, never mind the heroic labours that followed in Gord's stories. After their pool sessions the students had progressed to learning in chilly lakes and off the nearby coast at Upnor, where a grim sixteenth-century artillery fort overlooks the murky waters of the River Medway's delta.

'They held my head underwater for a minute,' said Gord. 'To make sure you could hold your breath that long.' It was a scary idea. Even if I could hold my breath for a minute underwater, it would surely feel much harder knowing someone was holding you down. Memories came back to me of school trips to the local swimming baths and the other boys forcing me below the surface. More than once, I believed I would die.

'But why?' I asked. 'I've heard that the number-one rule of scuba diving is you must never hold your breath.'

'Yeah, that's what they tell you.'

'So why did they make you do it?'

'Because if you run out of air underwater, you can't find your buddy and you need to swim to the surface, it comes in bloody handy. Know what I mean? That's why PADI make you swim a certain distance while holding your breath. You've got to be able to do it.'

In that moment, I decided to moderate my learnt concepts with the inner voice of Gord. If I could summon the courage to try diving, then I would try to keep his attitude stowed in my pocket. Gord never accepted anything he was told without first understanding the rationale behind it. He wasn't a contrarian sceptic: he just knew that people and systems are fallible and that you need to understand what you're doing, the better to keep yourself and those close to you safe. He would curse sailor-fashion, but it seemed justified. The curses were a fuel, a means to drive away stupidity; a hammer to strike home the crucial lessons that might mean the difference between life and death.

If I got in the water, Gord's voice would be with me. I needed it to counterbalance the voice of Rán. The problem was the fear. For every waking vision of beautiful mysteries in the water, I had a nightmare of entanglement in the dark heart of the sea. Could a person overcome these things and learn to dive? When Gord spoke of times he had been afraid, it was always for a specific reason. They were things that would have given anyone a shock. It wasn't the water itself that frightened him.

I thought back to my exit from the dive shop as Gord and I stayed up late talking. He opened a bottle of Spanish brandy.

'I've been thinking about trying to learn to dive myself, Gord.'

There was a brief silence.

'Yeah?'

'Yeah.'

Quiet again. A flickering hint of a smile.

We talked on as the clock neared midnight. It became clear that for Gord and his family, diving went far beyond the status of a mere hobby. It was long-standing and generational, like farming or a serious heritage of sport. Trips abroad had frequently revolved around diving.

'Any chance we got, me and Jack would jump in the car and drive down to Estartit. Catalunya. He had a BMW, a straight six. We'd do the drive in a day. Couple of stops, that's all. Lot of good diving down there. We took the boys with us. They got started early. Dean got a bit further along than me. He's a rescue diver. Yeah.'

Gord's son Dean had trained to make controlled ascents in treacherous conditions while encumbered with life-size, weighted human dummies. Perhaps this was an unremarkable act for a diver, but to me it was astounding, a hallmark of competence, and of a bravery I could not be sure I possessed.

I felt a challenge in these exploits: a dare to try it myself. It turned out I wasn't the first one.

'My old man started later in life. I'll tell you what happened. He came to our club one night, he was in his fifties then. He said he was interested in diving. Some of the fellas laughed. *Have you heard? Gord and Jack's old man's on about diving! Ha!* Well he didn't like that. He gave up smoking. Started running. Couple of months, he was fitter than us. He did his training, passed. He's in his nineties now but he was still diving until a few years ago. Credit to him. They said he couldn't do it. So he did.'

Gord's family had a range of motives for diving. Compared to the things I sought from the sea, what they returned with was less nebulous. Often, it bore blood.

'We always took a bag. Never know what you'll find. Edible crab, maybe a lobster. Or we'd have the odd flatfish, plaice, sole. Or a skate. If you're quick.'

'How did you get them?'

Gord produced a huge black-handled knife. It looked like a prop from a film about the Vietnam War.

'I've read that no one carries those massive knives any more,' I said.

'Why not?' asked Gord.

'They say they're not actually useful when you're diving.'

Gord shrugged. 'Looked pretty useful to me when I stuck it through a two-foot plaice.'

From what I'd learnt, many in the world of scuba diving would be appalled at these words. The dive community had largely turned against taking anything out of the water, whether living or not. But Gord and his family would never pass up a free meal. They were feeding themselves with the life of the seabed, as hunter-gatherers had. It seemed unfair to hold a person in contempt for doing something so ancient and human, just because civilisation had left us bereft of the need or the bravery to do it any more. And on top of this, a diver taking single animals by hand must rank among the least ecologically damaging ways to get food out of the sea.

'No different to my lot catching rabbits years ago,' I said.

'Not really,' said Gord.

As Gord talked on, some of the less obvious dangers of diving cropped up.

'What often happens in the winter is your regulator freezes,' he said. 'The water's cold and you've got cold air coming in and it just ices up. You want to watch out for that: it's how a lot of people have cocked up. And you only cock up once.'

I took a calming swallow of brandy.

'I tell you what will put the wind up you: getting caught in a school of mackerel. I wouldn't say it's quite like a load of knives hitting you, but it did shock me the first time. You just get clobbered. It can go on for ages as well.'

I quivered at the thought of trying to keep composed in a maelstrom of fish: shards of muscle, each one lancing through the water at 'burst speeds' of up to eighteen feet per second. It's difficult to find estimates for how many fish might make up a large mackerel school, as there are often so many in one place that they get measured in terms of distances rather than numbers of individuals. Atlantic mackerel can migrate in dense groups more than five miles long.

'And of course, the sharks come after them. So there's that.'

I let out a nervous cough. Gord decided to change tack.

'Anyway. I've got a wetsuit in the loft,' he said. 'It was Jack's. He was about your height.' His voice cracked a little. 'It might fit you. Maybe you can try it and see how you get on.'

In the twenty-first century, the wetsuit had become a mundane item. But there was something about the way Gord had stowed away these old garments, storied habits of the sea hidden behind a wooden hatch in the dark, that was redolent of mythology. It brought to mind the tales of the selkies: legendary beings of Celtic and Norse coastal mythology who, by using an enchanted sealskin, could shift their shape from seal to human and back again, living awhile on land, awhile in the waves. The magical skins were real: Gord was going to get one out of his attic.

'It might be a bit perished now. Might not be no good. But it's worth a go, ain't it?'

Gord climbed the steep ladder up into the roof space and shone his torch around. The small bulb cast a glow like that of a candle, and in it the beams and trusses of the loft resembled the bowels and struts of an old ship. He reached inside and pulled something towards himself. In the darkness it didn't look like a manufactured suit. It looked like a hide. A skin.

He hesitated. Then he said, 'Right. Here we go.'

Gord inspected the suit. Time had parched it. He ran his fingers along the scuffed seams, riven apart in places, where thick stitches ran through the neoprene. Around the ankle cuffs and the crooks of the arms and legs, its wrinkles and splits looked almost sore, injured like the cracked skin of an elephant seal. It had the aura of an artefact, a relic, a vestige of great deeds. It didn't matter that it was only a few decades old. It was a mummy of an oceanic life. It had been dipped in sea-time.

Gord decided it was probably too far gone, too long out of the water. But it was a talisman. To handle this suit, this second skin that a man I'd known had donned on underwater adventures, sent fresh belief surging through me. Jack had worn that suit in the kinds of places few humans ever go, pushing his cylinder ahead of him until narrow passageways of rock opened into glittering caves, their ceilings hung with crystals millennia old. The reality of these feats was commemorated by the brittle uniform of the sea in which Jack had done them. Now I had touched the hem of his garment. I felt transformed.

It was time to go and try for myself. To get closer to Dad; to get beyond the limits of what he would have done, of what he would have considered safe; to find out if the sea could tell me things about Atlantis, that crazy old word, my obsession, which I would never be able to learn here, on dry land.

But will it work? Or will you fail?

Might you . . . die?

I kept silent in front of Gord. These were the questions I had no answers to, and neither would he.

The First Breath

Orihuela Costa, Spain

The light of the sun was tracing gentle beams across my eyes, and the air rushed into my lungs, cool and pure as winter wind. A miracle. I was breathing while submerged. I breathed out again, and took a more powerful breath. Happiness surged into my body with the cold air. I could not drown.

I had a trip to Spain coming up. Candis's grandfather Lars was Norwegian, born in the far north of the country where the sun sets in November and does not return until the 'polar night' ends in the new year. He went to sea as a teenager, worked for a time on a whaling ship, became a marine engineer and never returned to his homeland. Lars had retired to Spain, and passed away a few months previously. He lived in the seaside town of Torrevieja, 'Old Tower', named for the watchtower of indeterminate date which stands near its northern beaches. We needed to sort out Lars's affairs and secure his house against would-be intruders. For me it also presented an opportunity. With its warm and relatively clear waters, the sea off the Costa Blanca would be an ideal place to put my dreams of scuba diving to the test. If I couldn't get through a couple of try dives in Spain, there was no way I'd handle the conditions back home in England. I would put my failure to sign up behind me, and start again. Walking around the house of an old man of the sea spurred me on.

I didn't want to go alone, though. I was hoping that Cand would come with me. I hid my fears and rhapsodised about how it would be glorious and was surely something everyone should try once in their life. She was unsure about it.

'I don't know. I'll have a think,' she said, with a slight frown.

I changed tack. 'Isn't it cool that you're from this . . . diving family?'

'What?'

'You know: Gord, Jack and Dean, your grandad.'

Cand's grandfather on her mother's side, George, was a naval diver in the wake of the Second World War. One of his jobs was to swim beneath vessels and inspect their hulls for damage. He had also worked as a mine-clearance diver, trained to diffuse waterborne bombs capable of destroying a battleship. Cand had shown me his navy diving log, with the rubber-stamped details of dives done at Aden and Suez during moments of international crises. I thought about Dad, an artist, and me, a writer, and felt in my chest the disquieting tug of difference between these ways of being alive.

'It's like a dynasty of divers. It's amazing.'

'That's weird,' said Cand.

'It's true.'

'I know. I suppose I just never looked at it like that. It was just . . . normal.'

'Normal!'

'Yeah. They'd just all done diving.'

'Exactly! Might be normal in your family. Look at mine. None of us has ever done it.'

Cand was thinking.

'Can't let the side down,' I said, with a stupid grin.

She shook her head. My plan was working.

I was buoyed by my conversations with Gord, and thoughts of him and his brother diving off the Spanish coast. I looked up the address of a local scuba operation and the following day at noon we headed there in a taxi which dropped us off into the brash midday heat. As the cab drove away, Cand's phone rang and she ambled off down the street, chatting. In front of me was the dive shop. I stepped off the searing pavement and looked in the window: diving posters, bits of technical equipment, and a selection of artefacts that looked like they had been fished out of the Mediterranean. There were clay handles broken off old amphorae, fragments of terracotta, and copper nails turned green by the sea. I went inside. The store had a marble floor and the cold seemed to seep through the soles of my flip-flops.

Suddenly there appeared a naked man. He was grey-eyed, with a steely and impassive face and sandy close-cropped hair. He had a lean, toned, freckled torso and was apparently wearing nothing but an enormous grey digital device strapped to his wrist. This was Kees Kouwenberg: Dutchman, expert scuba instructor and the resident 'divemaster' of the town. He said nothing, just stepped out from behind the counter. It transpired he wasn't nude, but clad from part way down his hips in a wetsuit half-unfurled, still damp. He started filling out a log sheet.

'Can I help you?' he said.

I identified myself as the Damian who had phoned earlier to enquire about a 'discover scuba' dive, the option for absolute beginners.

'When were you thinking? It's possible. But perhaps it will be difficult. The calendar is quite full.'

Cand had finished her phone call and she walked into the shop.

'Hi,' he said. 'Were you interested in diving too?'

'Maybe,' said Cand.

'Well if it's both of you, I can take you out first thing in the morning!'

I smiled.

'The first dive is a pool session,' he said. 'We meet at the pool and I show you how everything works. And we can check that you will be OK in the water. If you are OK in the pool, then we go to the sea. OK?'

He made an OK sign, touching the finger and thumb of his right hand, and flicked it towards me, as though throwing a dart.

*

The pool turned out to be located in the grounds of an outdoor bar. My romantic images of scuba diving — gaunt mariners with sheathed knives strapped to their legs — were entirely absent from the tableau before me. British, Dutch and German holidaymakers sizzled on spring-loaded plastic loungers, which made a carping sound as people sat down. Children wearing inflatable armbands ran around squirting each other with water pistols and throwing pieces of party food. Speakers cackled out 1970s Europop. This, then, was where we would learn.

Kees was there, a lone figure of concentration, standing apart from the tipsy holiday scene, surrounded by gas tanks, big goggles, snorkels, wetsuits and flippers. There were also the stab-vest things, with toggles and buckles hanging from them. These, I later learnt, were called BCDs — buoyancy control devices, or buoyancy compensators. They allow the diver to achieve neutral buoyancy when underwater — to hang in space like an astronaut in zero gravity — and to establish positive buoyancy when they need to float at the surface. Divers require them because they lack swim bladders or other means to control their buoyancy: because they aren't fish.

The puzzle that has driven the evolution of fish is not oxygen specifically, but air in general. Oxygen can exist in solution in water, and enter marine organisms by passing across their cell membranes. Air — in the form of actual bubbles of gas — is a different matter. Air is problematic for sea-creatures because its volume changes greatly depending on the atmospheric pressure of the water; hence, at depth, the risks for an organism with pockets of air in its body are too great. Deep-sea fish, therefore, do not have swim bladders like their shallower-dwelling pelagic (upper, open-water) relations, and instead control their movement by other means. Some use dynamic lift, which in simple terms means constantly swimming. Sharks are a well-known example. Their overall density is higher than that of the surrounding water and so they must sink or swim. Other fish also (or only) attain neutral buoyancy by using stores of fats or oils in their bodies, fluids with densities lower than that of seawater. Since these liquids, unlike air, are incompressible, their buoyancy does not alter with depth. A liquid means of control seems like an elegant solution to a biophysical problem – certainly more elegant than the assemblage of tank, valves and pipes required by scuba divers to achieve a poor imitation of a fish's underwater stability. Many common fish – including sharks and rays – are cartilaginous and lack swim bladders. So do the chimaeras, which are largely deep-water species with wide eyes and a spaced-out look, known variously as spook fish, ratfish, rabbit fish and ghost sharks.

Bar the odd manufacturer's logo or slice of colour, Kees's BCDs were all black. They resembled part of a superhero's armour, as opposed to something that would actually keep you afloat in choppy seas. In my mind's eye, that would be a standard, over-the-head 'Mae West' type life vest; the kind worn by Allied fighter pilots during the Second World War, hopefully in some garish shade of fluorescent yellow or orange so that rescuers would have a chance of seeing us against the blue, green or black of the depths. There was no sign of those.

Kees started putting things together and telling us what they were. Nothing was called what I thought it was called. The air tanks were 'cylinders', the big goggles were 'masks', the flippers were 'fins', and his watch, apparently, was a 'computer'. It didn't look like a computer, and I couldn't believe it was necessary – or wise – to strap a computer on your arm to go underwater. Clearly this was going to be even less like swimming or snorkelling than I'd imagined. How had I managed to learn so little from the conversations with Gord and Jamie? Maybe they'd held back on the vocabulary because it would only have confused me. The one thing I recognised was the regulator, because it had a rubber mouthpiece attached to it that looked like a boxer's gumshield.

It turned out there were actually several regulators, and they all had different names. They are called regulators because they contain valves that regulate the pressure of the air that comes through them.

The air in a scuba cylinder is stored at high pressures, typically between 50 and 200 'bar', one bar being the atmospheric pressure of air at sea level. The big valve that screwed on to the cylinder was a 'first-stage regulator'. It's the first stage the air passes through on its journey from cylinder to diver, and it brings the pressure of the air down to the ambient pressure of the air or water surrounding it. This is crucial because a diver needs to breathe gas delivered at the same pressure as their surroundings, otherwise they risk serious injury. The breathing mouthpiece was a 'second stage', and the spare, yellow breathing mouthpiece was an 'octopus', which was particularly confusing, since unlike the first-stage regulator, which had several hoses sticking out of it, this 'octopus' seemed to have only one arm.

Kees gave us another look up and down. He handed us a wetsuit each and sent us to the changing rooms to put them on. I had never worn a full-length wetsuit before and I felt a little squeamish stepping into it. It dawned on me how many people had probably worn it before, and that not all of them would have waited until they were back ashore to relieve themselves.

Then there was the inescapable fact that wetsuits have a lot in common, sartorially, with latex fetish gear. In my imagination, a wetsuit was a magical second skin, a svelte, blubber-like hide that enabled a human to be a little more like a marine mammal. But actually putting one on, I couldn't decide which it was more like: a selkie-skin, a superhero's costume, or a gimp suit.

We walked back over to the pool. Kees glanced at us quickly. 'Ah, they fit,' he said. 'Good.' He clearly had no interest in whether either of us looked cool. He must have seen thousands of people mooching about awkwardly in hired wetsuits during his life as a diving instructor, and he had more important things to worry about.

'OK. Let's get started,' said Kees.

He got started by describing in rich detail the worst possible things that could happen to someone while diving. Seizures, burst arteries, excruciating back pain, incontinence, nausea, paralysis, brain malfunction, the sudden appearance of scars with no visible cause. He must have noticed the terrified expressions on our faces, because he started to row back a little, reassuring us that all these nightmarish things were unlikely to happen. As we were total novices, he was only allowed to take us down to thirty feet. Diving at such modest depths for relatively brief spans of time, we needn't worry about these dangers as they are all associated with decompression illness, colloquially called 'the bends'. Cand and I looked at each other in relief, before Kees began to list the things that can still go wrong in shallow-water diving: exploding lungs, burst eardrums, entanglement, and drowning: i.e., death. But the main cause of all these was panic, and that wouldn't happen because, as long as we remained calm, there would be nothing to panic about.

I looked at Cand for reassurance, but the blood seemed to be draining from her face.

Fortunately there was an end to these chilling matters. We had to learn not just the dangers, but also some skills if we were going to actually dive. Kees went through the basics of diving. He started with how to equalise, to ensure the pressure in the air spaces inside your ears was equal to the pressure of the water outside. Without doing this, you couldn't dive below six feet or so or you'd risk doing irreparable damage to the inner ear. Kees demonstrated three methods: swallowing, wiggling the jaw from side to side, and the Valsalva manoeuvre. The last is named after an Italian anatomist born in the seventeenth century, and is the most popular method of alleviating pressure in the ears on an aeroplane. You squeeze your nose and keep your mouth closed, and then try to breathe out, which adds air to the space behind the eardrum. People tend to prefer one of these methods, Kees said, but if we had any problems it was best to come up in the water slightly and keep patiently trying all three of them. He told us, using his hands, how to 'fin' – or kick your legs – properly, which should be done from the hips and not from the knees, to conserve energy and generate more propulsive power. He showed us how to clear any water that managed to get into our masks, which was done by looking up towards the surface, pressing the top edge of the mask, and breathing out forcefully through your nose, which would fill the mask with air and push any water out from beneath the eyes. If our regulator or, at the surface, our snorkels got full of water, we could clear that by blowing out through them hard. I liked the idea of being in an environment where you could solve so many problems using nothing but your breath.

It was a lot to absorb. But we needn't worry too much, because Kees was going to demonstrate it all while the three of us knelt on the floor of the swimming pool. He'd do all the 'skills' and we would copy them. When he was confident we could do them properly, he'd give us the OK sign. Then we'd swim around the pool for a bit to get the hang of operating the jacket, which I was getting used to calling the BCD. And crucially, we'd have to accustom ourselves to the surreal fact that we could now breathe underwater.

Kees assembled the equipment on us, dressing us in it like we were children going to school. When it came to the sea, this was exactly what we were: marine neophytes who needed constant supervision. It was a relief that he helped us get kitted up, because it all seemed indescribably complex. And heavy. The downward pull of the cylinder, and of the lead blocks on my hips, was intimidating.

'Wow,' Cand said. 'It weighs a ton, doesn't it.'

'Yes. But you get used to this. Eventually you hardly notice,' said Kees.

He told us to inflate our jackets by pressing the red button on the tube that hung down by our left shoulders. The jackets made a satisfying hissing sound, quickly swelling up like a blood-pressure armband. This was somewhat restrictive to movement and breathing, but in return you got a powerful sense of being protected by a kind of air-filled armour. 'Yes, it makes a very good life jacket,' said Kees. 'You see how you don't need to worry?'

Kees double-checked our tanks were fully on, and then it was time to finally enter the pool. We would get in at the deep end because we were going to use the 'giant stride' entry technique, which is where you stand on the edge of the pool and take a long step forward to make sure your cylinder clears the edge as you drop. Kees told us to remember not to just walk in anyhow. The pressure of the inflated jacket, the sound of excess air hissing out, and the tremendous weight of the equipment made me feel like a living puppet, a human special effect, rather than a selkie or merman. That beautiful holistic harmony and unity of an underwater animal was not to be ours. Kees showed us how to use one hand to hold our mask and regulator in place, fingers outstretched as if splatting a custard pie into our own faces.

Now it was time to put on our fins. 'You don't put them on until as late as possible. No walking around in them like an idiot on the cartoons. You cannot afford to fall over whilst in your equipment. If you do, you will not easily get back up.' Bending over carefully so as not to lose our balance, we tugged the ludicrous flippers on to our booted feet. Once they were in place, though, they felt oddly satisfying. It was as if we really did have marine limbs, long things with a latent power to propel us in the water, to waft and swoop. I felt a new empathy for penguins and seals, beings that look ridiculous on land. They must itch to get in the sea where their forms are better suited, as I did now.

Kees stood on the edge of the pool, checked his gauges and the computer on his wrist, clutched his mask to his face and his BCD to his body, and made a long stride forward. For the briefest moment he hung in the air, then with a thunk and splash he sank into the pool.

Almost immediately afterward he was back at the surface, and he touched one hand on the top of his head and called out, 'When you are OK, you do this. To tell everyone else that you are OK. OK?'

'OK.' It had almost stopped sounding like a real word.

'Now you get in. One at a time.'

I stood on the precipice and held everything tight to my body. There were a mere six feet between my eyes and the shining surface of the water, but my body knew it was strapped in weights and encumbrances and would not let me believe that it would float. My muscles tensed.

'Come on,' said Kees. I drew a cold blast from the regulator. And I stepped forward into the air.

There was the familiar smash of liquid and confusion, but because of the mask, it came without the usual surge of chlorinated bubbles forcing their way up into my nose. I hung below the surface, tugged down by the cylinder and the belt of lead. This was it: down to the bottom I would go. I braced myself. But then something happened. Before I got there, the momentum was killed, and my body surged back in the opposite direction. I was yanked upwards as though by an invisible hand. It was the air in the BCD. My head surged through the water and into the light. The equipment, the physics, had won. I could not sink. We had been proofed against drowning. I felt augmented. I touched a hand to the top of my head, as Kees had done. I was indeed OK.

Cand strode in next. She made a more graceful job of it, and hardly seemed to sink down at all before she was back on an even keel. She shook the fizz from her head and touched her scalp: she was also OK.

Kees beckoned us towards him. 'Now we will go down.' He told us we'd kneel on the bottom of the pool and practise the skills we'd discussed. We held the inflator hoses up, let out the air that was keeping us afloat, and slowly, like human submarines, we started to sink below.

Our heads were under the water now. I looked up. Above me the surface was silver and furrowed with light, an alien sky.

Instinctively my arms got into position to pull me back up towards the air. But then something happened. I remembered the regulator was in my mouth. I stopped moving, exhaled through it, saw the blossom of bubbles surge away, and I drew a long, cold breath.

I will never forget that first breath underwater. I wanted to tell everyone in the world that they had to experience this. Imagine living your life and never knowing this feeling. Imagine never experiencing the thrill of this transgression, against the laws of lungs and liquid.

I sank further. I saw Cand with one hand on her inflator hose and the other hand pinching her nose, and Kees, perfectly still in the water, as if harnessed in it, keeping an eye on us both. He beckoned us towards him, used the OK sign to ask if we were all right, then swept one hand up his torso and back in a Zen-like gesture that intuitively meant 'breathe', but also something more: 'Contemplate your breathing. Enjoy your breathing.' I breathed slowly and deeply. He nodded his head.

Kees gestured at us in turn and made a different sign, a downward fist with the index and middle fingers of his right hand bent halfway and tucked in so they resembled a pair of kneeling legs. Then he pointed at the floor.

I did the Valsalva manoeuvre to equalise, and heard the tiny squeak of the air moving out through my Eustachian tubes. Those little pipes inside the head, of which we are usually so unaware, became suddenly present, as real as my eyes. The three of us assembled on the mosaic-tiled floor of the pool, kneeling in a triangle like underwater monks. One by one, Kees demonstrated essential diving skills, and we did our best to copy him. He took the regulator out of his mouth and threw it away to the side, as though contemptuous of the fear of drowning. He then leant down to his right, stretched his arm out and swept it around in a wide arc, found the regulator and placed it calmly back into his mouth. He blew through it forcefully to clear the water, then breathed in. We copied, less elegantly but well enough to get the OK sign from Kees. Next, he lifted the bottom of his mask slightly away from his face and let it fill up halfway with water. He looked skyward, pressed the top of the mask, and breathed out through his nose. The silver waterline slid down the inside of his mask until it was filled with air again. He held palm-up hands towards us, as though politely opening a door, inviting us to do the same. This skill felt less intuitive, and after a brief moment where the water was leaking into my nose and I wanted to get out of the pool, I felt the dryness return and was content to remain where I was. Surprised and relieved that it worked, I watched Cand do it next. She seemed a little agitated by the process, but did it well.

Now it was time to learn how to clear our mask if it became fully flooded. Kees took his mask off and knelt before us. He was squinting through the chlorinated water. He stretched the rubber band back over his head and secured the mask in place, still squinting at us through the water that filled it. He did the same thing as before, looking up and pressing the top edge of the mask while breathing out through his nose. We watched as the air pocket with its shining silver meniscus expanded, pushing out the water, which was followed by a few bubbles that crept from below Kees's eyes. He made it look like a magical power: as if he was expelling the water by an act of will, a silent spell.

Taking the mask off made it harder to believe we'd be able to keep breathing through the regulator. The unconscious mind finds it difficult to tease out all these processes, especially when they are new and unfamiliar. My breathing became irregular as I pulled the mask back over my head. In front of me, Kees was an unfocused shape gesticulating instructions that were tricky to see, so I acted from memory. I looked up and snorted some air out of my nose. It wasn't enough: the mask was still full of water. In my peripheral vision I could just about make out Kees doing the 'breathe in deeply' gesture, summoning fullness towards his ribcage with his hands. I did as he indicated, then let out a much longer and fuller breath from the nose, and my mask was now half-clear, my eyes in the dry. Encouraged by this, I did it again, and the mask was empty. Kees gave me a double OK sign, one with each hand. I was impressed it had worked. Once Cand had done it, Kees gave us another sign: he used the index and middle fingers of his right hand to mimic a pair of legs kicking, and told us to go and swim off around the pool.

For ten minutes or so we swam about in the narrow space between the floor of the pool and the top of the water. It was tough to obtain the feeling of weightlessness that divers often speak about: I either seemed to be rising slightly, or sinking. When I looked across the pool to Cand it seemed that she was experiencing the same thing. The blocking of the nostrils by the mask created a feeling of not being able to breathe. I dealt with this by occasionally snorting air out of my nose to reassure myself that I could breathe out through it, at least. But we were swimming underwater, breathing underwater, living underwater, and we had a rough equilibrium that made it all feel gentle and effortless, relaxing and serene.

By the time Kees signalled for us to surface, I was convinced I would do this for the rest of my life at every opportunity, until I was too old to do it any more. Even in the swimming pool of a hotel, it felt like we had experienced a scientific marvel crossed with watery witchcraft. We got out of the pool and undressed. Kees told us not to worry about the kit – he'd sort it out – and reminded us when we needed to be back in the morning for our try dive in the sea.

We rode back to the house in a taxi, and I felt the afterglow of having experienced something new and special: the loss of a naivety that I hadn't even realised I possessed.

'I didn't like the mask,' said Cand. 'I couldn't get rid of the feeling of not being able to breathe.'

'Me neither,' I said. 'But it's worth it, I reckon.'

I wondered if we'd be able to get over that and enjoy the sea-dive in spite of it. I forced doubts to the back of my mind. We had taken the first crucial steps. I felt full of light and hope. Perhaps I was really going to turn into a diver, a human who could breathe beneath the sea.

Atlantis Since Plato

Auschwitz; Byzantium; the 'New World'

June, 1944. A teenage boy's parents are sent to Auschwitz. The boy hides out in his grandmother's house in the Drôme *département* of south-eastern France, where he passes his days in fear for his life and for the lives of his mother and father. He also reads books. He is obsessed with the *Iliad*, the tale of the siege of Troy which had helped to supply a sense of cultural unity to the often quarrelling peoples of ancient Greece. The quarrels of his time are deadlier than anything the poets of Troy or Achaea could have imagined.

A year passes, the war ends. The boy's parents, who went through the gates of Auschwitz and never returned, do not survive it, but the boy does. His name is Pierre Vidal-Naquet. He will go on to become one of the modern world's great classicists. Though he also writes about other topics, including the harrowing events of the Algerian War, his central passion is the world of antiquity. He has a special corner of his heart where the idea of Atlantis lives. And he has his reasons.

★

One thing we can say for certain about Atlantis, that most tricksy of words, is that its history starts with the Greek philosopher Plato in the fourth century BC, just shy of 2,400 years ago. There is no record of a place called Atlantis before Plato mentioned *Atlantis nesos* – the 'island of Atlas' – in his writing. Supposed mentions of Atlantis by writers prior to Plato are, without exception, references to other people and places. Hesiod, in his *Theogony*, uses the phrase *Atlantis Maië*, but this means 'Maia, daughter of Atlas' and is not a reference to a place. Herodotus, writing in the generation before Plato, again uses the word 'Atlantis' but with the sense of the Atlantic Ocean, not an island. The proto-historian Hellanicus of Lesbos, a contemporary of Herodotus and the first non-Roman writer to mention the founding of Rome, also used the word 'Atlantis' with the meaning 'of Atlas': in this case it meant, as in Hesiod, a daughter of Atlas the Titan, not Atlas the king of Atlantis. All subsequent mentions of Atlantis the place, whether in later Greek and Roman literature or above the doors of kebab shops, can be traced back to Plato.

The idea that Atlantis was first mentioned by a philosopher might sound intimidating, but Plato was not just a thinker. He was a writer, and one of the greatest writers of all time. His ideas have inspired so many later minds that the British mathematician A. N. Whitehead referred to the entire European tradition of philosophy as 'a series of footnotes to Plato'. From his thoughts about how the ideal country should be run, to the famous allegory of the cave, and his theory of the 'forms' that lay behind the human ability to interpret the world, few thinkers have ever been as influential. Plato's greatest gift to popular culture, however, was none of these things. That gift, unquestionably, was Atlantis.

Plato wrote in dialogues. His writings all feature multiple people talking to each other. If he were writing now, we might call him a playwright, but one whose works were read off the page instead of staged or filmed. Calling Plato an outright dramatist is an oversimplification, however. As the renowned translator Benjamin Jowett observed, 'We lose the better half of [Plato] when we regard his Dialogues merely as literary compositions.' There is much in the dialogues that truly happened, and some things that happened are disguised as fiction.

Whatever we want to call Plato, he probably wrote in this

way – with different people talking to each other – because the mind is inherently dialogic. Humans are social animals, evolved to live in groups, and our brains find it easier to follow a conversation than a long monologue: we prefer a chat to a rant. We should remember, whenever we mention Atlantis, that it first came up in a script, a highly confected mode of writing that had more in common with a play than a straight historical narrative.

Any serious book about Atlantis will mention that the story first occurs in two works by Plato, the *Timaeus* and the *Critias*, both thought to have been written around the year 360 BC. The dialogue of *Timaeus* is the longer of the two – almost 33,000 words in translation, of which about a sixth are connected to the Atlantis story. And within this section, only a 400-word sliver directly describes the lost island kingdom.

By contrast, Plato's dialogue of *Critias* is mostly about Atlantis. The *Timaeus* introduces the mighty island, but the *Critias* is Plato's only 'Atlantis book'. For a work by Plato, it is very short: the length of a long magazine article. In English translation, just under 7,000 words of it 'survive'. That's if we believe that the end was lost, as many have assumed it to be. It might never have existed at all. Plato may have deliberately abandoned it: perhaps even broken off mid-sentence on purpose in order to make it feel more real; to give the reader that sense of reading a scroll that is suddenly, thrillingly, torn at the crucial point.

The *Critias* is the source for many things that people still think about Atlantis today. It establishes the idea of Atlantis thriving a long time ago and being situated in the Atlantic Ocean: 'in all 9,000 years have elapsed . . . since the war between those who lived outside and those who lived inside the Pillars of Heracles.' It describes the Atlanteans' vast wealth, which 'was greater than that possessed by any previous dynasty of kings or likely to be any accumulated by any later'. It details their technical sophistication: 'this may sound incredible for an artificial structure,' says Plato, describing the enormous canal dug by engineers around the island's plain. The account ends with the sudden destruction of Atlantis, which 'was subsequently overwhelmed by earthquakes'.

Not everything mentioned in the *Critias* is quite as enduring. The military power of Atlantis has proven less interesting to popular culture, but it was of major concern to Plato. He says the island was able to supply up to 60,000 officers – or 'leaders of military detachments' – and that it had an 'unlimited supply of men in the mountains and other parts of the country'. It could field 10,000 chariots, and had 'a complement of 1,200 ships'. To put these numbers in perspective, modern historians estimate that the vast Persian army which invaded Greece under Xerxes I consisted of about 800 ships and perhaps 150,000 soldiers (far short of the 2.6 million claimed by Herodotus). With 60,000 senior soldiers each leading his own detachment, Atlantis's army would have dwarfed even that of the Achaemenid Persian Empire, in its day the largest empire the world had ever seen.

Then there are the many things in the text which rarely crop up in modern iterations of Atlantis – like gods. The *Critias* says that the island of Atlantis was sculpted by the sea-god Poseidon – 'turned as if with a lathe and blade' until it was to his liking. He then 'fathered five sets of male twins' with the human woman Cleito and set them up as kings over ten different parts of the isle. This origin story sounds bizarre to us, but to the ancient Greeks it would not have. Powerful dynasties usually traced their lineage back to a sexual union between a mortal and a god, and the *Critias* also mentions how the primeval Athenians themselves were descended from Athena, the goddess of wisdom, and Hephaestus, the blacksmith of the gods.

The bits about 'primeval' Athens are the most commonly forgotten parts of the *Critias*. Atlantis-hunters tend to overlook them; Athens is real, we might reckon, and therefore boring. But the Athens in the *Critias* is not the Athens of history. It is Plato's idea of a 'prehistoric Athens', of many thousands of years ago, and it contrasts with the Athens in which Plato lived almost as much as it differs from Atlantis. In those very ancient days, we are told, Athens was not a coastal city: it lay much further from the shore, and possessed rich and fertile farmland. It had nearby mountain forests which supplied good timber for building. The Acropolis was thickly covered in soil. The citizens had no private property. 'The military class lived by itself . . . surrounded by a single wall like the garden of a single house.' In short this Athens would have been unrecognisable to the Athenians of Plato's time. By then it had become 'a mere remnant of what it was', 'rather like the skeleton of a body wasted by disease'. Atlantis may have been destroyed beneath the sea, but Athens was also a ruin of its former glory.

In the years after Plato's death, a debate quickly began as to whether the Atlantis story should be taken at face value. Proclus, a philosopher writing about 800 years after Plato in the fifth century AD, tells us that the question divided Plato's Academy, and classical thinkers in general, into different camps. Aristotle and Eratosthenes were suspicious of the story, but others (including Crantor, who became a leader of the Academy, and in later generations the geographer Strabo and the polymath Posidonius) thought it might well be true. Were they idiots? Or were they on to something?

★

The historian Tom Holland observes that the idea of Atlantis we have now is 'at least as modern as it is ancient'.[1] It is impossible for us to think of it without the centuries-worth of images and concepts that encrust it like barnacles on a wreck. In many ways, the original story and the various structures that people have built on it are now inseparable, melded together into a new entirety. Atlantis now resembles a tel, a settlement containing layers and layers of remains, built one on top of the other over long spans of time, until it becomes a hill composed of history. Or we might see it as a palimpsest, a waxen book in which words have been etched on top of earlier writings, sometimes many times over, until the original text is so obscured that it becomes impossible to make out. In such things the old and the new become fused.

The reason Atlantis has become so layered is because there is something precious in it. The tel is reused because it has a well of fresh water, or commands the view of its surroundings, or sits amid fertile ground. The palimpsest is reused because parchment or papyrus were in limited supply, so old words had to suffer the new to lay upon them. The idea of Atlantis has great value, and so we layer things on to it, time after time, because it is better than trying to find a new word for the purpose.

In spite of this, it bided a long time almost completely ignored. For over a millennium, from the fall of the western Roman Empire until the middle of the fifteenth century, Atlantis was rarely referred to. This is partly because all of Plato's writings apart from the *Timaeus* were lost to the West in the Middle Ages, and it was the rediscovery of Greek literature during the Renaissance that led to the re-emergence of Plato's dialogues in Latin Christendom. This was followed by a resurgence of interest in the Atlantis story in the sixteenth century, by which time it was almost 2,000 years old.

What caused this rediscovery? Scholars largely agree that a key factor was the westward migration of the learned classes from what had been the Byzantine Empire, following the fall of its capital, Constantinople (now the Turkish city of Istanbul), to the Islamic Ottoman Empire. These émigrés brought with them texts and traditions of scholarship that had survived in the eastern half of the Roman Empire, many of them transmitted from the classical Greek period in which Plato lived and wrote. Although the Italian Renaissance is generally agreed to have begun earlier, and as early as AD 1300 in the view of some historians,[2] it was this later influx of old Greek learning that helped trigger the birth of Renaissance humanism and played a key role in the development of modern sciences. In the Atlantis story, the Egyptian priest at Sais had lamented that there was no such thing as an ancient Greek, but by the time of the Renaissance, there was nothing so revered as classical Greek learning, of which the Atlantis allegory was part. Plato would surely have enjoyed the fact that his story, about a lost maritime empire rich in arcane knowledge, was delivered to the future by mariners from a real lost empire rich in arcane knowledge. (The author and philologist J. R. R. Tolkien also relished this sort of notion: sailors fleeing disaster, bringing mysterious histories and secret sciences with them to new realms, play a key part in his created mythology.) There is an irony in the mistaken popular theory that Atlantis was a real place that spawned great civilisations. Athenian culture, via the Renaissance, had a central role in the generation of the modern age. So the empire that seeded later cultures was not Atlantis, but Athens, the city of Atlantis's creator; the city which, in Plato's story, had been Atlantis's arch-enemy.

*

In the fifteenth century something else was afoot that would catapult Atlantis back into cultural relevance. This was the dawn of the imperial age of Europe. The discovery that the Atlantic Ocean didn't go on forever – that there were, in fact, lands and complex civilisations on the other side – was bound to jog the European memory of the Atlantis story. Crucially, Plato's tale contained the idea that the powers across the ocean were the bad guys. There seemed to be a moral warrant in it: 'kingdom across the Atlantic bad, Europe's warriors good'. This was to prove convenient when it turned out that the 'New World' contained material treasures beyond the dreams of Europe's rulers, alongside the less convenient fact that hundreds of indigenous cultures already lived there.

These newly encountered lands could not be the Atlantis of Plato's writings, of course, because that had long since been destroyed. Still, it was tempting to wonder – had they inspired Plato's tale or been part of Atlantis's empire? Might it have stretched not only eastward across the Mediterranean but also into secret, rich lands in the west?

For some, these sorts of ideas proved irresistible. The belief in Atlantis played a part in the conquest of the Americas and in Hispanic colonialism, influenced by the thinking of the Dutchman Jan Gerartsen van Gorp, also known as Goropius Becanus, and the Spanish polymath Pedro Sarmiento de Gamboa. In his *Historia general llamada índica*, Sarmiento de Gamboa claimed that Atlantis had once adjoined Spain, that America was the remains of Atlantis, and that America therefore belonged by right to the monarchs of Spain. A key figure in this matter was Francisco López de Gómara, a Spanish priest and historian writing in the middle of the sixteenth century. López de Gómara never went to the Americas himself; but he did spend time with conquistadors, including Hernán Cortés, when they returned to Europe. López de Gómara listened to the conquistadors' stories of what they had seen in Mesoamerica. The Aztec city of Tenochtitlan, with its systems of bridges and causeways, its floating gardens, encircling waters and mountainous vista beyond, must have sounded tantalisingly like Plato's description of Poseidon's defensive systems of moats: the water-embroidered capital of a powerful empire far to the west. In 1552 López de Gómara became the first European to declare that the New World was, in fact, Atlantis. Never mind Plato's assertion that Atlantis had been completely destroyed 11,000 years ago: López de Gómara's claim required that the complex histories of indigenous American peoples, and the methods by which they were recorded, could not be taken seriously, and that in some cases they must be systematically ignored.[3]

As the European states' colonial ventures intensified, so too did the willingness of some writers to justify them by using even wilder riffs on the Atlantis story. In 1673 another Spaniard, José Pellicer de Ossau y Tovar, became the first to identify the site of Atlantis with Andalusia in southern Spain, where the kingdom of Tartessos had supposedly been located. This theory would mean that the Spanish royal family, as descendants of the kings of Atlantis, could lay claim to its former empire. To us it is a preposterous idea, but the tethering of Spanish monarchs to the invented kings of Atlantis was to play an important role in the colonial thinking of the Early Modern Period.

Of course, this narrative was not just totally incompatible with Plato's writing – which made it clear that Atlantis was not situated in mainland Europe, but was a giant island in the Atlantic Ocean – it also contradicted the assertions of López de Gómara, who had declared that the New World *was* Atlantis. Yet, in their own ways, each justified the Spanish colonial project. Both writers took advantage of Plato's prestige to provide a pretext for colonialism. This only worked because two things were true. The first was that there was such immense respect for the great Athenian philosophers that Plato's writing could be cited almost like holy scripture. The second was that there was such a powerful sense of ethno-national grandeur in the Atlantis story: of great deeds done by nations of the past. Plato had painted a picture of a world in which tremendous feats on a mythical scale were achieved by archetypal peoples vying against each other, a tempting lens through which to view the colonial age.

Spain wasn't the only place where the Atlantis story was used as a pretext for empire building. Olof Rudbeck was a seventeenth-century Swedish academic who argued in his book *Atlantica* that what had once been Atlantis was now Sweden. He claimed that Scandinavia had been the first region settled by humans after the great flood of the Bible, and that a magnificent civilisation grew up there, which survived in cultural memory by various names, including Atlantis. Rudbeck's theorising was part of a hurried attempt to seize hold of mythologies that would lend cultural heft to the designs of Swedish imperialists. If their country was going to rule the entire Baltic region, to believe in itself and be accepted as a resplendent overlord, it would be helpful to root its legend in an equally glorious past.

Decades later, after the Swedish Empire had largely crumbled, the Piedmontese thinker Giuseppe Bartoli put forth what is now the consensus view among classicists, which is that the Atlantis of Plato's story was a symbolic, fictional empire. It represented the Athens of Plato's time, and of which Plato disapproved: a power-hungry state with a fearsome navy, with echoes of the Persian Empire that had menaced Greece a generation before. The resonances in Bartoli's argument, of the now-humbled Swedish state, would have been obvious in his time, but they came too late to save Rudbeck from the madness of his own theory.

These visions of Atlantis were not just dreams, or experiments in creative writing. They were ideas: influential notions that ultimately contributed to ethnic cleansing, industrial slavery and the racial categorisation of humankind. Often they contained strange quirks which had little or no connection to Plato's writing. But ideas have a reality of their own, containing the power to change the world, to liberate, to subjugate or destroy.

★

Pierre Vidal-Naquet loved Atlantis, but the boy who survived the Nazis had no time for the literalists who thought that Plato's story was an account of a single historical place. His calm, bespectacled expression hid a ferocious appetite for information, and a determination to defend Plato's work against the wild theories of the dilettantes. Slapdash or wishful attitudes to the Atlantis story only served to cheapen its brilliance, and they could be dangerous. For Vidal-Naquet, it was immoral to value a romantic yearning for origins more than the riches of the text itself, riches that sit there waiting to be read by anyone who can be bothered. In his seminal 1992 essay 'Atlantis and the Nations', he addressed the hypothesis that the Greek island of Santorini – known in antiquity as Thera – had been the 'real' inspiration behind Plato's tale. 'This theory . . . is as absurd as all the others,' he wrote. Like most esteemed scholars who have analysed attempts to find a single, real Atlantis behind the story, Vidal-Naquet believed that all literalist interpretations of it were equally misguided, each one stemming from a fundamental misunderstanding of Plato, whether wilful or naive. Why else would they ignore so much of the text? They overlooked the copulation of the god Poseidon with the human woman Cleito, and the idea of primeval Athens being a republic, millennia before such a political system had been conceived of. What of the other details in the descriptive passages, such as the anachronistic presence of hoplites and triremes, symbols of war in Plato's time, not the Stone Age? And what about the swords? The oldest swords ever found, made from arsenical bronze, date to around 3300 BC and were unearthed at the mound of Arslantepe in the Taurus Mountains of Turkey, a site hundreds of miles from the sea: surely an Atlantean sword would have turned up somewhere by now? What about the fact that large land masses do not sink overnight, even if cities can, and have?

Vidal-Naquet's position, of course, grew out of a lifetime's study of classical antiquity, but it is also possible that his family's experiences at the hands of the Nazis played a role in his thinking. The proper study of ancient texts could yield insight, nuance and joy, where irrational obsessions with ethnic roots had left Europe drowning in blood. Leading Nazis were enamoured of the Atlantis story. A key figure in the background of their fixation was the racist academic and pamphleteer Herman Wirth. Wirth circulated materials that promoted 'supra-rational magic' over the critical study of history and literature. Together with Wirth, Heinrich Himmler founded the 'Study Group for Spiritual History' called *Deutsches Ahnenerbe* (German Heritage), an SS think-tank. The idea of Aryan descent – in its pseudo-historical, Nazi sense – from Atlantis was popular among its members. It was the *Ahnenerbe* that, among other abominations, commissioned August Hirt, Rudolf Brandt and Wolfram Sievers to assemble the collection of Jewish skulls that was to be showcased at the University of Strasbourg in order to demonstrate the supposed inferiority of non-Aryan races.

Atlantis's appeal to the Nazis, alongside its employment by apologists for Spanish colonialism, has given it a lasting taint. In the eyes of some, the story cannot be washed clean of the fact that it just so happens to serve the interests of evildoers so brilliantly. We cannot blame Plato for the way his invented empire has been deployed and abused throughout history, but we might ask whether there is something in the story that has earned it the love of so many who would trample over other cultures. We might draw a comparison with how British imperialists justified their conquests by comparing themselves to the Romans, or the Nazis' admiration for the ideals of ancient Sparta: it was partly the brutal ethos of these bygone cultures that made them so beloved of their modern equivalents. For this reason, many intellectuals have entirely lost their appetite for discussing Atlantis, because past conversations about it have led down such hideous paths.

Vidal-Naquet himself argued that it wasn't the story that was the problem: it was the credulity of its readers and wielders, who ignored its historical context and most of what it actually said. A powerful dimension of the Atlantis story is that it is an anti-imperial text,[4] aimed at the warmongering disaster-makers of Athens in Plato's day. If there is an eerie foreshadowing of future ethnic strife in the story, it is the fact that it references distinct nations – Atlanteans, Athenians, Egyptians – at all. This is scant grounds for condemning it as a pretext for genocide.

While any wholly literal reading of Plato's tale of Atlantis is demonstrably absurd, it doesn't follow that every attempt to link the story to concrete historical happenings is equally silly. Among those who have looked for the pieces of a historical Atlantis, there may be a 'hierarchy of absurdity'. Nothing is ever invented out of a total vacuum. When Heinrich Schliemann and later archaeologists uncovered the remains of ancient Troy at the tel of Hisarlik in modern Turkey, they may not have solved the mystery of the creative process behind Homer's *Iliad*, but what they did show was that it contained at least a kernel of historical truth. If, as some believe, the works of Homer were the product of many successive minds, this would echo the structure of the site of Troy itself, a single mound composed of the layered remnants of many lives: a palimpsest of brick and bone. In the case of Atlantis, there is plenty of scope for such kernels of story: inspirations that are absolutely real, tangible ruins behind an invented world.

The Trace

Cala Cortina, Spain

We were twenty feet or so beneath the surface, perhaps the height of a small house. The skin of the ocean rippled above us, a curtain of restless splendour. Suddenly it looked like the frontier between this alien world, and our home. So this was it. We were breathing underwater. Out of our element, but alive.

We were to drive to Cala Cortina, a small bay on the eastern side of the harbour of Cartagena. Kees insisted on leaving at dawn and when we got there I realised why. A small road snaked down to the shore. At the waterside there was a beachfront café that smelt of bleach and vinegar. Beside it was enough room for four or five vehicles: we seized the best spot in the corner and within minutes the others had been taken. As the sun climbed the sky, a stream of young men in souped-up cars began snarling into the car park. Upon seeing that it was full, they swore from their windows in Spanish while executing five-point manoeuvres to get back out.

Basking in our victory, I smirked. Kees did not. He always seemed to wear the same expression of chiselled calm.

I went into the café and was met by a woman with a bronze sheen to her face and a look of power, a matriarchal air. I smiled at her. She did not smile. I pointed in the direction of the toilet and attempted Spanish.

'*Lo siento . . . pero, puedo?*'

She shrugged in reply, and wandered back behind the counter. Maybe this was one of those places where everyone used the conveniences but not everyone bought a drink, and resolved to return for refreshments after we finished our dive. Outside, Kees had opened the back doors of the van and was poring over his equipment. A saline breeze rolled in off the glinting bay.

Cartagena enters the historical record as a colony of the Carthaginian Empire. Its name comes from the Latin *Carthago Nova* – New Carthage – and since Carthage itself was called *Qart Hadasht*, the Phoenician for 'New Town', Cartagena's name means 'New New Town': ironic, given that it has been continuously inhabited for 2,250 years. Carthage rose from a mere Phoenician colony, situated on a hook of what is now the Libyan coast, to become the chief rival of the rising Roman Republic. However, the Carthaginians differed from Rome in that they had no aspirations to possess great tracts of the earth, rarely venturing into the interiors of the lands whose coasts they settled. Instead they targeted strategically vital ports and strips of shoreline backed by fertile land, using these to amass great wealth by a mixture of cultural influence, naval supremacy and trade. A map of Carthaginian territory can appear underwhelming, but the clue as to why is contained in the word 'territory', which comes from the Latin root *terra*, meaning 'earth'. Carthage was preoccupied not with the earth, but the water. So you have to adjust your eyes and imagine the sea itself as the arena, and the land as merely footholds: launch-pads of sorts. The true shade of their empire was the blue of the Mediterranean.

There were powerful cultures in Spain long before the Phoenicians. Cartagena is the possible – though disputed – location of the old port city of Mastia, and as such it might have once marked an eastern outpost of the fabled Tartessian culture, a semi-mythical civilisation that centred on what is now the Gulf of Cadiz and the Guadalquivir River. In classical times the Guadalquivir was navigable from its mouth as far inland as Cordoba, and to this day it is the only river in Spain that can accommodate ships. The Greek historian Ephorus of Cyme refers to 'a very prosperous market called Tartessos, with much tin carried by river, as well as gold and copper from Celtic lands'. Tin may not sound exciting but it is an ingredient of bronze: without it, there would have been no Bronze Age. Some have identified Tartessos with the biblical Tarshish, and the German archaeologist Adolf Schulten believed it to be the source of the Atlantis legend. Herodotus mentions a king of Tartessos called Arganthonios, whose name looks suspiciously like it means 'man who had lots of silver'. It is hard to know how much store we should put in old mentions of Tartessos; these stories could be etiological, invented long after the fact to explain the nature of a place. In any case, no one has ever found a definite trace of Tartessos. Schulten's quests were in vain.

We stood around in the car park as a scratchy radio blared from the café. I watched Kees busily adjusting shiny bolts and gauges with a selection of metal spanners and Allen keys: he was a man who had lots of silvery things. It turned out he wasn't just a scuba diver: he was also a mechanic, a van driver, a first-aid responder, a businessman who ran his own dive centre where he trained not just divers but other instructors, a fluent speaker of at least three languages, and his torso looked like it had been sculpted by Praxiteles. He was a sort of Renaissance merman. Suddenly I felt like I had paid good money to stand here feeling clueless in his shadow. Cartagena may or may not have once been home to a legendary water-going culture, but today it had Kees.

It was finally time to get ready. I grabbed my belt of lead weights and started trying to put it on. It was tricky. 'The best way is like this,' said Kees. He bent over forwards and slung the belt over his back, then did the buckle up beneath him. I followed suit: it was much easier, and an early lesson in how, with diving, a little knowledge saves a lot of work.

It was disorienting to feel all the weight on my hips: only my legs were affected, while my torso was free as before. A nylon strap full of lead: it made an odd alternative to the belts of gold sovereigns that Special Forces soldiers have supposedly worn in case they need bargaining chips in enemy territory. In our case we were bargaining with the sea, negotiating a way to sink; to settle in.

Then we had to put on our BCDs. With the air cylinders now attached, they looked like jet packs ready for take-off. Kees lifted them up in turn and helped us get them on. Again he told us to bend over until we had wiggled them into place and done up the Velcro 'cummerbunds' round our waists, and the several plastic clips. It had been a long time since anyone had helped me get my arm through an armhole. It was like being babied.

It was time to assess our seaworthiness. Kees looked us over as though we were show dogs, frowning and grimacing, tugging at things that were too loose, re-clipping twisted buckles, while we broiled in our suits. It was shocking to think that people did this as a leisure pursuit, and absurd that some of them even seemed to believe it was glamorous. The weight of the tank was pulling me backwards so to counteract it I leant forwards slightly, with rubber tubes sticking out haphazardly, numb flailing tentacles. My awkwardness was heightened by the sight of Kees buckling his weights on coolly like he was donning a pistol holster, and effortlessly swinging his BCD on to his back. His coordination made a mockery of the idea that bulky muscles are necessary to be strong. Somehow the equipment seemed half the size on him, as though made to measure, a natural adornment. He looked streamlined. Prepared. Marine.

'OK. So we go in the sea,' said Kees.

★

We trudged off down to the shore. People slick with tanning oil were draped along stripy beach towels. They looked at us with a range of expressions: bemusement, indifference, intrigue. I felt twice my usual weight. Every underfoot pebble pressed cruelly into my soles. Candis walked on carefully, saying nothing; Kees seemed positively jaunty. We hadn't even got in the sea yet. It had never occurred to me how much of scuba diving could happen before you had touched the water.

When we got to the tideline I became nervous about getting the equipment wet. This made no sense: it was scuba-diving gear after all, purpose-built for immersion. But I had only ever swum in the sea in nothing but a pair of trunks or, occasionally, less. Sea-swimming was a liberation from encumbrance; from 'stuff'. Back home, some swimmers wore rubber caps, goggles and gloves, which enabled them to stay in for longer but also removed, from my perspective, a portion of the charm. To be coated by the water, entirely in contact with it, was something magical, even if in cold temperatures it could only be briefly enjoyed. Feeling the rush of the water over my skin had become something close to a fetish. It felt wrong walking into the Mediterranean rigged with appliances, like carrying luggage into the sea. I couldn't shake the idea that something was going to short-circuit or explode. It was refreshing, though, when the water seeped into our boots and crept up under the anklets of the wetsuits, rinsing the heat from our exhausted legs. Once we were in chest-high water, the heaviness of the belts and cylinder suddenly disappeared. In seconds, the awful sensation of seething heat, and feeling like we were laden with enough weight to take us plummeting down to the bottom, was transformed into blissful coolness while we floated about whimsically, pulling on our plastic fins.

Kees told us we'd soon descend. We were to try and stick by him. 'The water looks pretty clear today so you should be able to manage,' he said. 'Maybe we will even see some little nudibranchs.'

'Nudibranchs?'

'Yes! The special little guys.'

Kees signalled thumbs-down and we started to let the air out of our BCDs. We were going down. Actually diving.

I tried to forget that my nose was blocked by the mask and to focus instead on the air that streamed into me through the mouthpiece. Though its coldness set my teeth on edge a little, it did help to calm my nerves, until my ears began to hurt. I squeezed my nose between thumb and forefinger and tried to breathe out. It didn't seem to be working and I was descending fast. Pain. Danger. I wiggled my jaw from side to side as Kees had also told us to do. That didn't seem to be working either. I swallowed. My ears gave a little mouse-like squeak and the pressure eased off just enough to let me calm down. I was desperate to be able to use my voice to communicate and the impossibility of doing so felt like a secondary form of suffocation. Kees gave no indication of that kind of distress. Above the water he was a man of few words, and under it, a man of none; but he managed to 'utter' his hand signals in such a way that the simplest gesture seemed full of nuance. He slowly lifted up a hand to give me the OK sign: it was obvious that it was both an instruction to remain calm and a question as to whether I was. I signalled back yes, I was OK, which was just about true. He did the same thing to Cand, and she signalled yes too. There was water in the bottom of her mask, but she seemed to be in control. I was putting effort into finning upwards to stay on Kees's level, and felt none of the serene sensation of weightlessness about which other divers rhapsodised.

I thought everyone would look slightly hapless underwater, but Kees did not. He exhibited a wizardly level of control over his movements, able to stop abruptly with no obvious sign of inertia, hoverfly fashion, and always looking relaxed with his hands clasped in front of him, a manatee-like posture, attentive, meditative. Occasionally he would spin round to make sure we were still alive, and he did so as though he had simply decided to in his mind and the rest of his body had followed suit without muscular effort. It looked less like swimming, more like levitation. At first I instinctively swam breaststroke-style with my hands, as I always did in the sea, but watching Kees move by using his fins alone made me feel clumsy, so I copied the clasped-hands style. I asked him about it later. He shrugged: 'It's good trim,' he said, before explaining that 'trim' is a diver's term for balance, posture, and correct positioning in the water. Besides being more streamlined, and allowing you to keep your instruments in view, it stops you using your hands to manoeuvre, which you wouldn't be able to do if you were holding a torch or navigating with a compass. Some divers with spinal injuries that prevent them kicking their legs do use their hands for propulsion, and to this end they wear specially adapted gloves which look somewhat like the hands of a platypus, with webbing between the fingers. But as a general principle, it seems hands are best kept folded away.

Occasionally I would forget that my nose was enclosed by the diving mask. I would try to inhale through my nostrils, and experience a moment of intense claustrophobia, a foul marriage of the blocked nostrils and the CO_2 that my body was desperate to expel.

We were near the bottom now. I was getting tired, perhaps as much from trying to convince myself I wasn't going to drown as from the actual exertion of swimming. I stopped briefly and stood still, suspended just above the seabed, the tips of my fins tickling the rocks. I was conscious of how even the slightest movement of my feet was dislodging a world, some arrangement of life that – as an animal only down here by the grace of human tech – I had no right to disturb. My 'trim' was non-existent, and a wayward swipe of a foot could send delicate flowers of seaweed surging upwards amid clouds of fine yellow sand. I did my best to avoid this, but when I failed it was awful – like traipsing over a field of wildflowers. Blooms of 'seaweed', which is actually marine algae that grows into structures resembling plants, were floating around me now like the heads of enchanted peonies. It was a signature of human blundering, but beautiful all the same. Nearby a column of black and silver fish, likely gilt-head bream, hung like a glittering mobile, keen eyes in every direction. In these fish, intention and motion seemed to be melded into a single act for which we have no word. If they were 'swimming' then there was such a contrast between their swift, chic, minimalist movements and my hapless churning through the water that they made the word 'swimming' seem unsatisfactory. Kees seemed to be somewhere closer to the fish than to me, on an imprecise scale of underwater finesse.

Then I looked up.

The immensity of the water, and the glinting blue-grey veil of the surface high above, gave a feeling of upward vertigo that no experience on land could have prepared me for. We were a mere thirty feet below, in a tiny bay of no particular significance, and yet the sheer amount of sea – the *space* of sea – was so enormous I felt close to the edge of a panic attack. Kees noticed and moved his flat hands up and down slowly. The meaning was obvious: *I know, I know. Stay calm. You'll be all right*. But my mind was racing away from me. Thirty feet, that's all: in places, the sea plummets to a depth of almost seven miles. The scale of the world was dawning on me, the majesty of its dimensions. My heart raced. I knew this feeling. I was remembering Rán, the terrifying power of the ocean. I thought about 'calling it' and doing a thumbs-up signal, which has a counterintuitive meaning underwater: the skyward-pointing thumb means 'end the dive'.

I was saved by an octopus: a real one. Kees waved at us and pointed it out, then made his calming gesture again to suggest a cautious approach. It refocused my mind. We swam over very slowly, side by side. There it was, hunkering under a shelf of rock, a great burgundy head with its many arms tucked behind, throbbing as it breathed the water. It was astonishingly well concealed, not just its colour but its texture mimicking the ruddy, encrusted rocks that lay around. And yet there seemed to be a pulsing aura of life that gave it away: perhaps they stand out more to our colour-calibrated human eyes, attuned for finding fruit amid the branches of primordial forests. We got within a few feet and it hadn't moved, a blob of Zen. It seemed genuinely to be meditating, sitting with all its arms tucked in, in a cloister between the seabed and the overhanging rock. Kees, in spite of his thousands of hours diving, was spellbound. His stillness communicated a reverence for this life, and for the moment: that this wasn't something you could take for granted. I was expecting the octopus to tire of our presence and retreat or become annoyed, but it simply waited. Outwaited us in fact, for our time was limited by the contents of the flasks of air on our backs. I wondered if it knew our time was short.

Nearby, on a fist of red rock, was a sea slug. The name sounds vile but this was a beautiful creature. It was lilac-coloured with flashes of flame yellow on its appendages, the stubby poisonous spikes that stuck out from its body. These have evolved as a form of protection, replacing the shells which its ancestors wore. After the dive, Kees would explain that this was one of the nudibranchs that he was hoping to see. Nudibranchs are some of the most colourful creatures on Earth and this one was no exception. Swaying from side to side in the current, it looked like a slug on its way to a festival or a rave.

Nudibranchs – whose name means 'naked gills' – lack the grisly sheen of terrestrial slugs: or rather, being underwater, their sliminess is not apparent. Maybe this explains why there are internet fan clubs entirely devoted to nudibranchs, while their landlubber cousins, with their tendency to shrivel up like vampires when surprised by sunshine, find it harder to win admirers. While the glamourless field slug curdles under a toddler's sprinkles of salt, the sea slug sways through the water column, posing on nature's catwalk.

Some nudibranchs make a beeline for the very animals we humans pray to avoid. They feed on hydroids – a class of marine invertebrates that includes jellyfish during their intermediate life-stage – and even highly poisonous adult jellyfish. Nudibranchs are able to process these animals' nematocysts – stinging cells – and arrange them near the surfaces of their own bodies. It is a form of stolen protection, a real-life biological equivalent to human cultures who have believed that by eating the bodies of their enemies, they might acquire their powers. Not even the most infamous hydrozoans are safe. The blue sea slug *Glaucus atlanticus*, also called the blue angel, sea swallow or blue sea dragon, is a top-dwelling nomad which specialises in feasting on the Portuguese man-o'-war. It floats belly-up on the skin of the ocean, magnetised to the surface tension of the water, bright blue and silver-white. As gaudy as a Kabuki performer, the ends of its cerata – long finger-like growths, arranged in fans – are constantly ready to venomise predators who trespass on to its stage.

We'd been underwater for forty-five minutes and started to swim back towards the shore. It was a relief to get back into shallower water. *Full fathom five thy father lies; of his bones are coral made*, sings the mischievous spirit Ariel in Shakespeare's play *The Tempest*. When we studied the text at school, I looked up how deep five fathoms was. Thirty feet, or just over nine metres. *Not very deep*, I thought at the time. *Could swim down there easily. Cheerfully undulate back up*. Shakespeare knew better: thirty feet of water – the depth we'd just reached on this dive – exerts a pressure almost as great as the entire planet's atmosphere above the surface.

We had entered the shallows, a depth of perhaps ten feet, when Kees spotted something. It was the remains of a white plastic bag. Haggard with the sea – it might have been down there years – it had turned a mottled cream colour and was ragged in places, torn into strips. It almost looked like a kind of pale seaweed, and I could see how a creature might mistake it for food. Kees swam over to it. He reached out to take it with him, but as he grabbed with a hand, the bag began to fall to pieces. Shreds of plastic slipped through his fingers and off into the sea. He gently gathered as much as he could, and put it into a pocket on his belt to dispose of later.

Currently at sea there are solutions to the colossal problem of human trash. As we swam, cutting the surface somewhere far away was the USS *Gerald R. Ford*, a gigantic American aircraft carrier of the latest generation. It employs a technology called plasma arc gasification (PAG) to turn all the solid waste generated on board into just two relatively benign substances. One is a high-value mixture of hydrogen and carbon monoxide called syngas, which can be used for everything from wax-making to iron and steel production. The other is a solid material called 'slag', a glassy rock-like substance which is non-toxic and can be used for building. There is a large initial investment required in order to start processing waste using this method. The energy requirements of superheating shredded household waste to 2,000 degrees Fahrenheit, a temperature at which all organic bonds are broken and metal turns to liquid, are intense. But once installed, PAG has the power to eliminate fundamental problems of other forms of human waste disposal, such as landfill, which almost inevitably leads to the infusion of 'leachate' – otherwise known as 'garbage soup' – into our planet's groundwater systems, causing the toxification of drinking water and in some instances the poisoning of local life. The idea that this foul system could be swapped for a process that produces nothing but useful gas and pellets of inert black rock seems too wondrous to ignore. Yet, as I write this, a gigantic ship is knifing through the surface of the ocean, able to turn its detritus into two things: one harmless, the other precious. Powered by nuclear engines, so as to have what is claimed by its makers to be an 'unlimited' range at sea, it is an unintentional riposte to the *Flying Dutchman* or the *Mary Celeste*: not a ghost ship, but a ship that lacks some of the characteristics of what we might naturally call a 'live' one. It does not require feeding in order to swim, and its very waste products bear little resemblance to the usual cast-offs of animal life.

At last the water was shallow enough for us to stand up with our heads at the surface. We leant backwards and took off our fins, then walked gingerly out of the sea. Emerging so slowly felt almost chic: as if we were rising by choice, rather than being flushed out by fear or necessity. But then the tremendous weight of all the lead and equipment re-announced itself. We were on land now, trussed up in encumbrances that didn't fit in this environment, and after the exertion of the dive it all felt twice as heavy. No wonder a beached whale is in such grave danger once it is stranded. That indomitable heft of muscle and blubber is not designed to lie there unsupported by the amnion of the ocean.

Hunched over, we made our way steadily back to shore. It was hard-going: our gear was laden with water, and we'd drifted northward a little on the dive. This meant the return route was longer than the walk down had been, and of course, it was back up the beach. We left wet footsteps on the path, and thin trails of water. The black of our saturated wetsuits glinted softly in the sun. We had returned alive from our baptism in Rán's world.

When we got to the van we unbuckled ourselves and let the gear drop slowly down to the floor. We were free. The sinews of my muscles seemed to breathe in relief at ditching the weight: everything expanded, our bodies lightened again. I took a deep lungful of air through my now-unblocked nose and stretched my arms up to the sky. Cand looked pleased to be out of the water, but more so, out of the kit. Then I glanced across at Kees. For the first time since I had met him, he was smiling. A beaming, uncontrollable grin under two little sparkling eyes.

'So. Did you enjoy that?' he asked.

It seemed less like a question directed at us, and more like an avowal of grace. The serious, fidgety, technical aura that Kees had worn in the shop, assembling equipment, making sure we were OK, was utterly gone; his silhouette was freer, expansive and changed. He looked like a man who had been to heaven, and knew he would go there again.

I went inside the café to buy us some beers and smiled at the stern proprietress in her starched clothes.

That evening I was full of fanciful ideas. I thought about how everything that divides our species above the surface seems to disappear underwater. Underwater we are animals who depend upon each other to share air if we're to survive. It supplies a vision of what life could be like if we simply switched our focus to other things. Astronauts who have seen the Earth from space often describe how this experience makes human divisions and conflicts seem absurd. Could the memory of having been to another world, the sea-world, render the notion of human divisions ridiculous and turn us all into siblings of the blue planet?

After nightfall our ears kept popping and crackling. We were scared. Had we damaged the fragile structures inside our heads? I sent Kees a long and panicky message. A reply came back.

Get some drops.

His nonchalance helped calm us down. He'd heard all this before from worried novices.

A trace of the salt scent of seawater clung to our hair, our necks and chests, sand-coloured and smooth like beaches swept by an ocean. A kiss supplied sensory evidence of the adventure: on each other's skins we could still taste a tang of the sea. We had slipped in and out of the ocean and escaped in our stolen cloaks, daubed with a paint invisible to the eye but plain to the understanding of the tongue. Perhaps the real sealskin was not the wetsuit, but this salt trace, this proof that you dared to get in.

The Puzzle

Marburg, Germany

Marburg, Germany; Christmas 1910. Alfred Wegener, a young academic and lecturer in physics, astronomy and meteorology at the university, was browsing through an atlas of the world.

Like every other atlas it depicted the 'old world' and the 'new', kept apart by a massive sash of blue: the continents of Europe and Africa, separated from the Americas by the Atlantic Ocean. Wegener stared at the map. He had seen it many times, but this time something told him to keep staring, to concentrate. He stared at the west coast of Africa and up at the coasts of western Europe, and he stared at the eastern seaboard of North America and down to the South American coast. He looked again. He felt something funny in his heart. And he looked again. It appeared that the east coast of the Americas would fit rather snugly up against the west coast of Europe and Africa: *would* fit, or perhaps *once had*? The thought of a jigsaw puzzle came to his mind. The pieces of a puzzle slot into one another's sides, because they had once been cut from a single whole. What if the land masses of Earth were like that: the pieces of a puzzle?

Wegener's pulse quickened a little. Was this one of those ideas too simple to have ever been noticed before this very moment? How could it be that he was thinking about it now; that he might be the first human to have ever had this thought? Staggered by the possibility that there might be something in it, Wegener closed the atlas. He would never be able to open it up with those familiar old feelings again. The pictures inside hadn't changed, but for Wegener, the atlas was altered forever. He would need to find proof, to test his theory in the fire of others' critical assessment. But if he was right, the world would never look the same. It wasn't just the sea that moved. The land was moving as well. There was a new liquid sense to the past.

We now understand that the history of the Earth's surface is a tale of the formation and fragmentation of supercontinents, of vast land masses accreting, then splitting apart. This is the deep past of every place. There was Pangaea – 'All of Mother Earth; All Land' – the name Wegener gave to the most recent supercontinent. It was a colossal mass of territory, which later split into two smaller – but still gigantic – sections: Gondwanaland in the south, and the Laurasian archipelago in the north. Over a period of 200 million years, both of these slowly broke apart until the world's surface gradually took on the form familiar to us today.

The restlessness of the Earth's continents works as a metaphor on a grand timescale for the way we experience the inherent instability of the world. Fragmentation always speaks of a previous unity – a prior integrity – interrupted and redefined. The ocean rushes in to form new edges, as water splits apart the blocks of the flooded city, or the bricks of the sunken house. All land used to be connected – now the sea lives in between.

Even if we have no primordial sense of this – if our feel for its drama comes solely from our knowledge of plate tectonics – it has still had a powerful effect on our psyches. We know that gigantic change has always been afoot: great turmoils of land and ocean, at scales that dwarf the destruction of a city, an island, a civilisation. Pangaea, during the Carboniferous period, was the biggest united land mass, which split apart during the Mesozoic Era. We might see it as a sort of slow, 200-million-year-long 'fall of Atlantis'. What if the idea of mighty forgotten civilisations, now broken and lost, has sprung in part from other notions of bygone unity and scale, unmatched in our own time?

★

Atlantis didn't only make its way into the narratives of fascists and apologists for colonialism. The story has also influenced artistic interpretations that are beautiful, as well as surreal and absurd. Pierre Vidal-Naquet divided the various recreations of Plato's Atlantis story into three general types. There was the 'pre-national Atlantis', a golden age which modern states could lay claim to as a legitimising force, prone to being co-opted by those hungry for conquests and power. There was also the 'cosmic' or Christian Atlantis, which included reimaginings of a Holy Land, or Atlantis as a spectral or metaphorical domain. Finally, there was the geographical Atlantis, which saw Plato's story as a cryptic reference to far-off places. In this view, the tale contained old mariners' expertise on the whereabouts of far-flung lands which had fallen out of common European knowledge. A 'geographical Atlantis' might be the Canary Islands, for instance, or the Americas. Some people's 'Atlantises' combined elements of more than one of these, but they all had one thing in common: they treated Plato's story as a true myth, something generated by a wider culture, and not as a fictional allegory which was the brainchild of a single author.

One of the first, and strangest, reworkings of the Atlantis story was created by a Christian traveller, and later hermit, from Alexandria in Egypt, called (in Greek) Cosmas Indicopleustès, or 'voyager of India'. In the sixth century AD, Cosmas made several trips to the subcontinent and his accounts contain some of the earliest known maps of the world. Cosmas attempted to weave Plato's tale into the biblical tradition, claiming that Atlantis was the garden of Eden, that Noah himself had lived there, and that the kings of Atlantis were in fact the ten generations of Hebrew patriarchs between Adam and Noah. In his view, this 'overlap' proved the truth of the Old Testament narrative, in spite of the fact that there is no discernible overlap between the two traditions at all. Indicopleustès's work, which also contained arguments for the Earth being not only flat, but square, has been called 'a monument of unconscious humour'.[1] What it does tell us, though, is that the reverence for Plato's writing was so strong that it was viewed not only as historically reliable, but capable of proving the truth of the Bible itself.

At first glance, the cosmic Atlantis would seem like the most harmless kind. Other visionary traditions have fed into this web of ideas around magical submerged places. Irish mythology tells of Tír na nÓg, an enchanted Land of Youth; the 'Celtic Otherworld', conceived of as an Atlantic realm 'without grief, without sorrow, without death'. Tír na nÓg overlaps with the Blessed Isles of Manannán Mac Lir, the Son of the Sea who rode his chariot over both waves and land. His three-spoked wheel may be the source of the triskelion symbol of the flag of the Isle of Man. Another Celtic tale tells of the monk Barinthus, who set sail into the west and landed upon the shores of paradise, and of his emulator, St Brendan the Navigator. According to tales dating back at least as far as the tenth century AD, Brendan, who was also called 'the Anchorite' and 'the Bold', discovered a string of magical islets on his voyage in the wake of Barinthus. He sailed in a craft of wattle covered by animal hides cured in the juices of oak bark and made supple and waterproof by the application of Irish butter.

Poets have tended to be drawn to the idea of a cosmic Atlantis as opposed to a historical one: an eternal, resplendent city, not the muddy remains of a seafaring empire shattered long ago. For them, Atlantis is an enchanted city with great symbolic power, lying mysteriously beyond the physical reach of explorers and navies. Far from being destroyed, it survives as a source of inspiration to voyagers, prophets and dreamers. The English poet and artist William Blake created an Atlantis that was very much of the psychedelic, technicolour kind. In his 1793 work *America a Prophecy*, he describes

> *those vast shady hills between America & Albion's shore,*
> *Now barr'd out by the Atlantic sea: call'd Atlantean hills*
> *Because from their bright summits you may pass to the Golden World*
> *An ancient palace, archetype of mighty Emperies,*
> *Rears its immortal pinnacles, built in the forest of God*
> *By Ariston the king of beauty for his stolen bride.*

Blake's conception of a glorious Atlantis, located in a distant stretch of the ocean far removed from Europe, was likely influenced by the work of Dante Alighieri. In his *Divine Comedy*, Dante wrote of an earthly paradise that lay a great distance away to the south and the west. It sat atop a giant mountain, the Mount of Purgatory, which was formed by the impact of Satan crashing into the heart of the planet. Satan's fall had caused an island to spring up at the antipode of Jerusalem: its exact opposite point on the other side of the world, an enchanted counterpart to the Holy City.

In his mythic reimagining of Britain, Blake mixes ingredients of the Atlantis story with Judaeo-Christian and Celtic elements. He turns the Hebrew patriarchs Noah and Abraham into druids. In Blake's story, Albion is originally the name not of a country, but of a giant: a son of the sea-god Poseidon who founds a kingdom on the island of Britain before being killed by the hero Heracles.[2] Prior to his death, Albion the giant spawns an entire race of giant children, who live on the island until the arrival of Brutus, a survivor of the fall of Troy. By the time Brutus arrives there are only a few giants left, survivors of lethal encounters among them. Brutus slays the last of the giants before setting himself up as king and supposedly naming the island after himself, Britain.[3] Over a thousand years later, Julius Caesar lands with a Roman army at Pegwell Bay in Kent. The ordinary history of the island of Britain has begun.

Blake worked to give Britain an Atlantean flavour, but in his poetry he also referred to Atlantis itself as a lost continent between Britain and America, two civilisations which were linked but mutually antagonising. This sunken 'Albion-Atlas' was a bridge between Britain and the New World, a place where 'Giants dwelt in Intellect' on the 'Atlantean hills', now lost beneath the water. Its hidden remains, though out of sight, still teemed with an age-old puissance. These were magical powers which, according to Blake, would occasionally rise from the depths to help steer the path of history, as they did in the eighteenth century when they influenced the American and French revolutions.

What if the 'cosmic Atlantis' and a physical Atlantis could somehow be combined? In the nineteenth century there was another surge of interest, and it was very different from those that had come before it. It may not be a coincidence that a new breed of Atlantis-seekers overlapped with the era of adventurer-archaeologists like Heinrich Schliemann, whose enthusiasm outmatched their subtlety:[4] they inspired lasting cinematic tropes of treasure-hunters who escaped the library to go gallivanting in foreign lands and smashing their way into tombs. This new 'Atlanteanism' blended a passion for lost or hidden pasts with aspects of science and attempts at geographical precision, and the Atlantis story played a key role in it. Atlantis started to be seen as the ultimate source of wisdom and culture, the wellspring from which all great learning had originally flowed.

Such ideas were connected with the belief that humanity was divided into distinct, discernible races. Where scuba diving fosters the sense of an international siblinghood of the water, the idea of descent from Atlantis has been put to more divisive uses. In some places, an Atlantean heritage was posited as an alternative source of the national knowledge base, for instance by the nineteenth-century Italian author Angelo Mazzoldi. Mazzoldi argued that Italy had once been the real Atlantis, and that the civilisation spread by Rome was therefore, ultimately, Atlantean. This theory would have been hilarious to Plato, since the culture spread by the Roman Empire was in large part Greek, and in his story Atlantis was the enemy of Greece. In spite of this, views like Mazzoldi's continue to prove popular. His work contained early traces of what is now called hyperdiffusionism. This is the belief that similarities between far-flung cultures – from things like pyramid-building and megalith construction to religious thoughts and rites – are best explained by a shared connection to a progenitor civilisation, a mighty parent culture that spawned them all.

To archaeologists, hyperdiffusionism reeks of quackery. They see it as a giveaway of cluelessness or pseudoscience. It is a tell that someone lacks an appetite for the subtleties of their discipline, its complexities and hard truths. Ironically, despite the ridicule they receive, hyperdiffusionists are seeking a cultural parallel to biological evolution – that humans share a common descent from a tiny group of ancestors. Hyperdiffusionism's adherents are drawn to the thought of separated peoples turning out to share a lineage from a powerful ancestral society. They often prefer this idea to other explanations which are seen as lacking its intuitive explanatory power. These include the convergent evolution of practices – customs in different places independently growing akin over many years – or of simple coincidence, stemming from the fact that humans are prone to doing similar things, even when separated by mountains, seas, and time.

Hyperdiffusionism doesn't always ally itself to thoughts of human unity. Some instances fit more easily into a hierarchical view of the peoples of the world, and are therefore ripe to be co-opted by racists: for instance, the idea that monumental architecture in the Americas must have been seeded by visitors from the old world, or even another one. Hyperdiffusionism has three big things going for it. It is simple, and therefore elegant, and it appeals to our innate sense of human connection. It presents the complexity of global cultures and their remarkable achievements as a puzzle with a single solution. In this it resembles the theory of plate tectonics, which helps to explain why the latter was initially ridiculed for offering too straightforward a solution to something as complicated as the layout of the world.

A hyperdiffusionist view could not arise, never mind enjoy the sort of popularity that it does, unless it chimed with a widespread inclination towards an ancient, lost, alluring oneness of things. Alfred Wegener could not have conceived of the notion of drifting continents in the first place unless at some level his mind accepted the idea of a pre-existing unity. It shouldn't surprise us that such an idea is latent in our spirit. Our own life's journey tracks a path from simplicity to complexity to fragmentation: from the single cells of the ovum and sperm to the adult human being, a being which in the end will splinter in body and mind until some new dispersal or mysterious 'second unity' is achieved.

In 1912, Alfred Wegener presented his theory of continental drift to the German Geological Association in Frankfurt. At first he got a muted response. Drafted into the German army during the First World War, he was wounded and used the time he spent convalescing to hone his ideas into a book, *Die Entstehung der Kontinente und Ozeane* (*The Origin of Continents and Oceans*). An English translation was published in 1922. The theory was dismissed as 'delirious ravings': 'Germanic pseudo-science' that took 'considerable liberties with our globe'. Wegener's detractors continued to champion their beliefs in geosynclinal theory and global cooling, claiming that mountain ranges had formed because layers of sediment got so heavy they weighed down the Earth's crust until it sagged in places, the effects of which were then amplified by the contraction of the planet as its temperature dropped.

As the progression of tectonic science in the twentieth century was to prove, it was not Wegener's ideas that were delirious ravings, but the beliefs that had preceded them. Plate tectonics, the more developed form of the continental drift theory, replaced them with a single, elegant, all-encompassing explanation of the facts. Not only had the wide Atlantic Ocean once been too narrow to host an island the size of Atlantis; if you went back far enough, there had been no such ocean at all.

Redemption

Wraysbury, England

The instructors absorbed the chaos with great patience. The contrast between their mastery of their every movement, and our struggles to keep still or move even vaguely as we intended, was staggering. They looked like adults beside toddlers, or even like a more evolved species of human, accustomed to life in the water. They didn't even look like they were in water. We looked like we were in water. They looked like they were hanging in space.

Back home, I walked through the door of the dive shop for the second time.

'Hello!' called a voice from a room beside me. 'Can I help?'

I looked in. There was a desk with a big computer on it, and nearby, a large aquarium full of little tropical sea-fish and anemones.

'Wow, that's beautiful,' I said, watching the tiny reef fish playing among corals that rippled gently in stripes of light. 'Are these yours?'

'Well there's a question! I . . . help look after them. I'm Angela.'

Something moved along the sand of the aquarium. 'What's that?'

'Ah, that's a little engineer goby,' said Angela. Her voice went up slightly in a manner that made it clear she was fond of the species. This one was a slender, worm-bodied fish with orca-like black-and-white patterned skin. Angela told me she'd reared it from birth, feeding it with a pipette until it grew in stature and bravery and eventually started to fend for itself. This one was still small – a couple of inches: older engineers can grow up to two feet in length, and they get their name from the tunnels they excavate in the rock of coastal reefs. A fully grown adult can move several pounds of sand from its burrows in a single day. They aren't just called engineers. They are engineers.

This felt like a very different entrance to last time. It had started by chance, but I liked how Angela had begun by talking about the habits of engineer gobies. It was a reminder of why people bothered to start diving in the first place. It was because there were things in the sea worth seeing and doing. I wanted that to come before the medical forms and health risks. Something about that little goby and Angela's knowledge about it had put me at ease. I felt like I was about to sign up for a bucolic undersea adventure, as opposed to a dangerous and technical extreme sport.

Angela stared at the fish. 'They're not really pets, but you do become attached to them,' she said. 'Fish are friends, not food.'

Friends. I felt a pang of shame at having eaten so many of them.

'Anyway, how can I help?' said Angela.

I explained that I wanted to do my basic open-water diver qualification.

'I'll introduce you to Jen,' she said. 'She can get you enrolled on the next course.'

Jen came out from the back room where all the cylinders and wetsuits were stored, as I explained I'd recently done a couple of try dives off the coast of Spain.

'And how was it?'

'Great.'

Jen smiled. 'Best thing you've ever done?'

I smiled back. 'Best thing I've ever done.'

'I remember my first try dive. Man. There's nothing like the first time! And it only gets better from there. Believe me.'

My grin widened. Jen's glee was infectious.

'Come on. Let's get you signed up.'

There was something familiar about Jen and Angela. It was reminiscent of the lady I'd met in the shop the first time I came in. It also put me in mind of Gord, of Kees out in Spain, and Jamie up in London. It was tricky to describe, but it must be something 'diverly', I thought. Jen and Angela looked like they could get on with stuff: fix things in a pinch, work through a problem, vault into the back of a truck. They wore dive-boat-branded polo shirts, jeans or cargo shorts, and 'buffs', a sort of elasticated neckerchief that doubles as a bandana or sweatband when required. They moved with purpose. They were too busy living to pose. And in every diver – even in Angela and Jen who I'd only just met – I'd witnessed a moment of sudden joy at the memory of the sea. It was characterised by the dawning of an unselfconscious smile, a mental return to the bliss of the water.

I wasn't sure I'd ever smiled quite like that. I wanted that smile.

Jen handed me the same form that had looked so intimidating the last time I came into the shop. This time I filled it out without much fuss: I'd had to tick these boxes out in Spain, and now that I'd had a brief taste of scuba diving, it felt less likely to end in disaster.

I was told to come back on Saturday morning with a pen, a T-shirt and swimming trunks, a towel, and some lunch. There would be two days in the classroom, followed by some sessions in the pool. Provided that all went OK, the following weekend we'd head up to Wraysbury to do our final training and qualifying dives in a lake there.

I'd signed up now; invested in the idea that I could be changed, made into something a little less of the land and a little more of the ocean. I was going to learn about sinking away from civilisation, slipping temporarily out of contact.

★

On Saturday I arrived early at the dive shop. There were five students; Lloyd, Philip and Dawn were about my age, and Jules was a boy of about twelve. His presence made me wonder whether I should have taken up scuba diving earlier – *imagine the experience I could have amassed by my thirties,* I thought – before I remembered how difficult that would have been while my dad was around. His constant worrying would have made it tough.

We were led into a narrow hallway redolent of a ship's corridor and full of the smell of fresh neoprene. On one side was a small, low-lit workshop suitable for a real-life Geppetto. 'That's Carl's room,' said Jen.

In there, bearded and focused, I caught a glimpse of Carl. He was assembling regulators, or possibly servicing old ones, removing from their metal parts the marine-coloured plaque of the Channel: salt crystals white as sea foam, and verdigris deposits with the mint and turquoise hue that was often, on bright but unsettled days, the colour of the sea nearby. The sea leaves its trace on everything, but its kiss must be cleansed from diving equipment before it corrodes it and impairs its functionality. I would learn that Carl was an accomplished technical diver with decades of experience. Nobody better understands the need to maintain the equipment than somebody who has depended on it for their survival at depths of 200 feet or more.

Technical diving takes place beyond the limits of normal recreational scuba and is often far more hazardous. A difficult term to define precisely, it generally refers to visiting greater depths than those for which most amateur divers are licensed, but it is also used in relation to penetration dives: the exploration of hazardous underwater environments such as cave systems and wrecks. Technical diving entails complex planning and generally higher stakes. One particular danger that must be contended with is gas toxicity: as pressure increases with depth, oxygen, the very element on which our lives depend, starts to become poisonous to the human body. This, in conjunction with the risk of decompression sickness presented by nitrogen exposure, means that divers working beyond 130 feet often use gas mixtures that contain a reduced oxygen content and no nitrogen at all: helium is used in place of the nitrogen, as it poses fewer hazards to the body under pressure. Such gas blends are known as 'heliox' for short, and common recipes include 84% helium with 16% oxygen, or 90% helium with 10% oxygen for deeper or longer dives. During their return to the surface, the technical diver will switch to different gas mixtures whilst carrying out their decompression stops: these will either be left hanging from a shotline, or carried by the diver throughout the dive. There is little room for error in these processes. The correct labelling and use of cylinders is a matter of life and death.

We went upstairs, past a couple of artillery shell casings that had gone green with sea-age, and into a room full of chairs, the walls decorated with stunning photographs of marine life: pufferfish, sharks, manta rays and exotic shrimps. There was a bar with a few bottles of spirits on the optic, and upon the dark wooden counter was a donations box for the Royal National Lifeboat Institution.

On the floor in front of the teacher's desk was a scuba cylinder, or rather half of one. It had been sawn in two lengthways to show the thickness of its steel walls. The space they contained was shockingly small. It was already hard to believe that dozens of cubic feet of air could be crammed into a cylinder, and that was based on looking at one from the outside. Seeing its tiny internal chamber made me realise anew how counterintuitive physics and engineering could be. Scuba tanks made it seem like space itself – the three dry dimensions in which we are used to living on land – could be shrunk down and made portable; smuggled into the ocean to help us feel more at home there.

Into the classroom walked two men. They were in their fifties, perhaps, and suntanned in spite of it being the tail end of a cold winter, bringing with them the atmosphere of a wild place far away. They wore shiny dive computers, black polo shirts embroidered with little flags – the white diagonal line through red that, at sea, means *divers below* – and name badges reading ANDY and MARTIN. Andy was slightly thicker-set, with cropped salt-and-pepper hair. His gaze, although friendly, lingered on you as if trying to figure out if you were sufficiently serious about the task at hand. Martin was wiry and bespectacled, quiet and observant, with sandy hair swept back like beach sedge in the wind. They hadn't said anything yet, but with their soldierly air and high-tech bracelets, they looked recently arrived from a distant land, a superior civilisation. They looked like real Atlanteans.

Andy announced that he'd be leading the course with Martin in support, and handed back the test sheets we'd done online beforehand, mumbling 'very good' to each student in turn, before he got to me.

'Up all night, were we?' he asked.

I looked at the paper. It had time stamps for when I'd done the tests. 'Not quite,' I said.

'Good,' said Andy.

Andy walked to the front of the class and asked us to say our names and a bit about our motivations for taking the course. Holidays were mentioned. Lloyd was embarking on a career change to marine biology. Philip was here 'just for the challenge'. I said I wanted to visit underwater ruins, so I needed to learn how to explore them without dying.

'Well you've come to the right place,' said Andy.

'Between us Martin and I have over fifty years' diving experience. We're into doing things properly. You can do this course abroad and there are places that will show you stuff once and tick some boxes and give you a licence. But that's not what we do. We teach people how to scuba-dive. We're here to help you to be safe. The qualification comes second. And you will not be learning to dive on some picturesque tropical reef with water at thirty Celsius and a lovely hundred feet of visibility. You will be in the pool downstairs. And then you will be in drysuits at Wraysbury, which is a green lake near Heathrow Airport. Where on a bad day you can't see your own hand in front of your face.'

We students exchanged alarmed glances.

'But we look at that as a positive. If you can dive there, you can dive anywhere. So by the time you have done this course, if you finish it, yes you will be certified divers. But more importantly, you will know that you know how to dive.'

None of us said anything. All of us were impressed.

'Of course it's also possible that some of you won't pass. You might get claustrophobic from the mask. You might panic. You might not be able to equalise your ears. Some people find it too hard to control their buoyancy. You might just not like it and that's fine. Diving isn't for everyone. But you only know whether that's you once you've had a go.'

We were silent again, but it felt a bit different this time.

For the next few hours we went through the theory we'd studied online. Basic atmospheric-pressure calculations, fundamental scuba-diving practices, safety. To remind us what can happen if you surface while holding your breath, we watched a video of a guy inflating a balloon at depth and then releasing it to the surface. It got about halfway, growing all the time, and then exploded. 'That could be your lungs,' said Andy. 'But it won't be, as long as you continually breathe as normal. In scuba diving you never hold your breath. *Ever*.'

The group had a few goes at assembling scuba apparatus. We made a mess of it. Tanks were strapped to BCDs with the valve pointing the wrong way; gauges were stared at in perplexity when they showed we had empty tanks: it was because we hadn't turned the tanks on. BCDs refused to inflate because we hadn't attached their hoses. It was a quiet comedy of ineptitude.

'Don't worry,' Andy said to me. 'You'll learn.'

★

After lunch it was time to take our gear downstairs and get in the pool.

Kneeling on the bottom, we repeated the exercises I had done in Spain: regulator removal and retrieval, clearing the mask, no-mask swim, and so forth. Andy and Martin elaborated on them. We signalled 'OUT OF AIR' by doing a karate-chop gesture across our own throats, as if to say 'off with my head'. Then our buddies offered us their octopus regulator and we finned up slowly, arm in arm, towards the surface. The 'buddy system' was growing on me. It was reassuring to know that there was somebody else besides yourself who was interested in your safety, and that you had their equipment to rely on if the worst thing happened and yours failed.

It all seemed to be going well until it was time to put on our drysuits.

The air temperature at the poolside was in the high twenties, and as soon as we started to don the suits it became clear what they were: impermeable waterproof shields, tailor-made chambers of discomfort which would trap heat and sweat – and us – inside. We wiggled our feet down into the integrated rubber boots, footwear suited to dangerous chemical work, or an abattoir. We squeezed our hands and heads through the latex seals – rubbery sphincters that grip the skin tightly to keep water out – then knelt down into a ball while pulling the neckpiece away from our throats to leave a gap. This expels any excess air. When we stood back up we were vacuum-packed. We stared at each other and raised our eyebrows. Did we look intrepid or totally ridiculous? Or both?

I took an instant dislike to the drysuit. This was not like a wetsuit at all: it gripped you with a tough, rigid embrace. It was hot. Next to our skin it took only moments for a microclimate to develop, the stickiness of a jungle. The drysuit forms an impermeable barrier between the diver and the water. It is a castle wall, designed to keep the lake's or ocean's touch off the vulnerable skin. I understood that it was impossible to scuba-dive in cold water without a drysuit, but I wasn't happy about it. Having often swum in cold water on purpose because of the health benefits and exhilaration it provides, there seemed to be something sad about this particular piece of equipment: it was a 'yes but no'; a symbol of entry but not of full embrace.

The drysuit had one thing going for it, though. It reminded you what diving really was: an audacious, and historically dangerous, act. You only had to look at it to know. Its appearance implied the entire perilous saga of diving's evolution. It was an imposing garment, the regalia of someone who had courage, or a serious job to do. Unlike a wetsuit, which people wear for many reasons – surfing, swimming, or simply to keep them warmer while they play about in the water – the drysuit with valves on means scuba diving and that's all it means. It reminds you to take the training seriously, because if it wasn't serious, then you wouldn't have to wear one of these.

As soon as we got in the water, a lot of the discomfort disappeared. The heat was drawn away from us and our burning skin calmed down, but there were new things to consider. The water pressure pushed the suits against our bodies, forcing the undergarments into hard ridges. It felt like invisible hands were squeezing my calves and thighs. This is known as the 'drysuit squeeze' and you alleviate it by pressing a button on your chest which adds air from your cylinder to the suit to offset the difference in pressure. I pressed the button and it gave out a quick hiss. The grip on my legs was instantly released.

But then I experienced the consequence of that relief. My legs were now more buoyant, and they seemed to want to drift upwards towards the surface, making control in the water more difficult. I also understood why we were wearing more lead weight than I had done in Spain. Our suits were full of air, not water, which made us much more buoyant. Drysuit diving is defined by the trade-offs created by this air: its insulating and pressure-relieving properties versus its buoyancy. The cost of keeping warm is that gas is always keen to get back to the surface. As a result, we were told to use the buttons on our chests with great caution, and to only add the minimum air necessary to offset the 'drysuit squeeze'. This was especially important when using a drysuit at depth, as adding too much air – air that would rapidly expand as you moved upwards through the water column – would increase the chance of a runaway ascent, with all the dangers that accompanied it. We were to keep a close eye on our buoyancy at all times, to ensure that our arm valves were kept partly open, and to remember to use the release button on them to let air out rapidly if we found ourselves rising fast. This felt like a lot to remember, given that we were also using our BCD to control our buoyancy. It was like learning to drive a car, thinking it was impossible that this tangle of observation and physical gestures could ever become second nature.

We repeated our skills in our drysuits, occasionally losing control and bobbing about or spinning around. We were joined by Angela now, which I found reassuring: perhaps it was connected to the way I'd seen her care about the fish in the aquarium. I was nervy in the drysuit and I wanted someone to care.

This meant there were now four novice student divers and three instructors all moving around within a relatively small cube of liquid. The divemasters somehow kept calm even when, in our throes of confusion, we occasionally kicked them in the head with our fins. They were aquanauts. I couldn't imagine ever being able to move that elegantly in the water.

At some point during the session, Jules had left the pool. I saw him later, wearing a dejected look. He'd had trouble equalising his ears and had to drop out of the course and go home.

'Some people are luckier with that than others,' said Andy. 'He's young. I'm sure he'll be back.'

I was full of admiration for the fact that Jules had even tried: at his age I would have been too afraid. I hoped he would try again.

★

The next week it was time to finish our training, and this meant we were going outside.

Wraysbury Reservoir is a vast, flat kidney of water containing 7.5 billion gallons: a number rendered meaningless to my brain by its sci-fi blend of size and specificity. Wraysbury supplies fresh water to London, siphoning it off from the Thames in enormous quantities every day. If the reservoir were to be poisoned or otherwise compromised, the metropolis would thirst, so its proper maintenance is critical for the hydration and sanity of the capital. A flock of sheep roams its borders, employed to keep the grass on its embankments short and tidy, making access easier for engineers. We were to dive in one of the small lakes west of the main reservoir, part of the local constellation of pools.

The lake's tranquil skin hides an improbable collection of objects. Most of them are vehicles, retired from service to slowly moulder away in the olive light. Beneath the grey flash of the surface there are twenty-eight sunken boats, a large bus, the fuselage of a 737 passenger aircraft, and the front half of an American taxi cab that once appeared in the feature film *Die Hard*.

The water was likely to be cool – twelve centigrade or so – and a dreary, tenebrous green. I was told the visibility at Wraysbury is not always bad, but that it would be today because of novice students 'standing upright like idiots' in the water, and 'kicking up all the silt and crap' from the bottom. This was a prophecy of my own behaviour, and a memorandum that it would be my own fault that I couldn't see anything.

My prescription for the lake's glum conditions was a drysuit, and beneath it a thermal undergarment that looked like a six-foot Baby-gro. I pointed out the resemblance and a fellow student wondered whether it might go beyond a similarity of tailoring. 'I bet some people shit themselves in these,' he said.

I was wearing the same training drysuit they had given me for the pool. It was the opposite of elegance. With its dayglo orange arms it seemed designed to make student divers look not just highly visible, but absurd, perhaps even dangerous: like traffic cones or poisonous frogs.

I put my weight belt on. There seemed to be two options for how to wear it: 'very tight', so that it simultaneously squeezes the belly and grinds lovelessly against the top edges of the pelvis, or 'extremely tight', so that it avoids the pelvis but strangles the midriff. I went for the 'very tight' option, reasoning that some discomfort on the hips would be marginally less distracting than being choked from under the ribcage.

We were briefed about the skills we would have to demonstrate – control of our movement in the suits, basic compass use, taking our BCDs and weight belts off and putting them back on again – and then we walked along the slippery pontoon and jumped in. In the soupy water there was visibility of perhaps six feet and no advance warning of anything. The supervising instructor was Angela who, like Andy and Martin, moved through the green of the lake with hypnagogic mastery. While we bumped and occasionally thrashed about, trying to wrestle control of our buoyancy and direction, Angela demonstrated a Butoh dancer's command of position and gesture.

In water like Wraysbury's, all things appear unheralded. The crayfish, the striped perch, the occasional hunting pike, the dive platforms, the lake bottom, and other divers, each of them jolting fears and excitements. Nothing is ever seen making its approach: they are absent from your world, or they are suddenly, shockingly in it, appearing with the instantaneity of characters in a dream. It can be panic-inducing. It forces you to live in the moment.

Half an hour spent in the company of inquisitive members of other species, in their world, made me ponder the way we go about describing them, and ourselves. There are many English words for things which are alive, and which live somewhere. Some of these words seem to only apply to humans: people, person, citizen, inmate, mortal.

There is aborigine, a word which is now used to refer to a member of an ethnic group who are indigenous to a specific land, but which originally had a wider meaning of anything alive that had deep roots in a particular region. Other words are more flexible: denizen, dweller, occupant, being; they encompass us alongside everything else that lives. Then there are the words that seem cold, an unsympathetic glance downward to those below humankind: sea-creatures, animals, beasts. Even 'living thing', with its implication of being an object, seems disrespectful. I wonder if I would call myself 'a living thing'; perhaps, but with a furrowed brow. I didn't want to use these cold words any more.

I completed the skills and an instructor invited me to buddy with her for a final dive near the middle of the lake. We would fin out and drop, then use our compasses to swim a basic square pattern for fifteen minutes before returning to the shore. At first the visibility was OK; we could see perhaps ten or twelve feet. As we finished off our planned pattern, though, it suddenly deteriorated to the point where we could barely see each other's faces when less than four feet apart. The instructor signalled me to end the dive. I copied the signal and gave the OK. Success. One dive down.

It was a mistake to think the dive was over before it was over, and I made a common, but serious, error. I held my inflator hose up to the surface so it would be in the right position to release air that expanded as we rose, but I released some air too early, for fear of becoming too buoyant. As a result, and unbeknown to me due to the poor visibility, I started sinking, in spite of the fact that I was finning upwards, trying to reach the surface. When I looked at my depth gauge I was confused to see that I'd dropped instead of rising. Because we were in the shallows, and I was sinking, my fin-kicks had sent clouds of particulate up from the bottom. The water was now opaque, and I could see nothing until I brought my computer inches in front of my mask. It told me I was now at twenty-three feet, ten feet below the depth at which we'd decided to end the dive. I couldn't see the instructor. I couldn't see anything. Even my sense of up and down was gone. I kept finning towards what I thought was the surface. I breathed a little harder. What was going on? Why the hell was I not at the surface? We were only in shallow water. How was it so easy to get so lost?

Suddenly the instructor appeared, her face close up to mine. I could see the relief in her eyes that she had found me: she must have surfaced and looked for my bubbles, and then followed them back down to their source. She used her inflator hose to indicate that I needed to use mine. I realised what was happening and the mistake that I had made. I put some air into my BCD and she pointed in the direction of the surface which, alarmingly, was not the direction I'd thought it was before. How had I managed to lose my sense of the vertical so fast? We stayed close together and finned our way up.

I was relieved to return to the light. Then I was told what had happened. 'We ended the dive and then you let air out of your BCD.'

'I know. I'm sorry.'

'Why did you do that? It's why you started to sink. And then I bet you exhausted yourself trying to fin back up while you were negatively buoyant.'

'I didn't want to go into an uncontrolled ascent. So I let out some air.'

'You need to start your ascent before you start letting air out. Otherwise you will sink. And in conditions like this it's worse. Because you can't see the surface. So if you get in a flap you could end up going in any direction.'

'I understand. I'm sorry.'

'Stop apologising. I'm not having a go at you. This is what training's for. I'm just letting you know what went wrong and how you can avoid it in future. Don't let the air out of your jacket or suit until you've started to move. Remember the procedure, the process . . . maintain control. Diving is great but you have to maintain control.'

When I came back ashore, Angela asked me and the other students to stand together for a photograph.

'Look,' she said to the other instructors. 'We've got ourselves some new open-water divers.'

The instructors clapped. They had done this many times but they could see what it meant to us, especially the ones who were afraid of the water.

Dad wouldn't have wanted me to do this, I thought. *But he'd still be proud.* For a moment I thought I might cry. But I smiled instead.

You're a diver. Maintain control.

New Questers

The Straits of the Mediterranean; the Deep Sea

Helena Blavatsky was a Russian mystic. She played the piano, had learnt some Tibetan from the Buddhist Kalmyk people of the northern Caucasus, and was an occasional bareback horse-rider in the circus. Charismatic and well travelled, Blavatsky was influenced by both Eastern and Western traditions in mysticism and philosophy, and fascinated by the idea of ancient lands that now lie under the ocean. In particular, she wondered about a lost continent called Lemuria, named by the zoologist Philip Sclater. Lemuria meant 'the land of the lemurs'. The distribution of these elegant primates had caused Sclater and others to wonder whether some primordial land mass had enabled their spread from the island of Madagascar to the Philippines[1] and beyond. In her 1877 book *Isis Unveiled*, Blavatsky touched on the idea of a lost Atlantic land. Her brief writings likely had a powerful influence on a contemporary of hers, who in 1882 published a book called *Atlantis: The Antediluvian World*. His name was Ignatius Donnelly.

Donnelly is perhaps the most important figure for understanding why so many theories about Atlantis have proliferated since the late nineteenth century. He was an Irish-American congressman who had lost all his money trying to set up a utopian farming community in south Minnesota, before reinventing himself as a bestselling author and theorist of human culture. Donnelly cast Atlantis as the world's first great civilisation, possessed of a genius that had never since been surpassed. It was the ancestor of the cultures of Africa, Europe and the Americas, the birthplace of metallurgy, of alphabets and medicine.[2] The great innovations of ancient times had been falsely attributed. For Donnelly, they all went back to Atlantis. The various regional cultures that were so proud of these contributions to human knowledge had not actually spawned them: they had merely funnelled them to us from their original source, the mighty long-lost island kingdom.

The idea of Atlantis the great ancestor culture has blossomed ever since. Due partly to Donnelly's work, Atlantis now acts as a lightning rod for those who challenge the consensus view of how Western civilisation was transmitted — who argue, for instance, that the part played by Greece was not as central as has traditionally been claimed in the West. The role of non-European cultures in the development of civilisation has often been colossally understated, from Mesopotamian and Egyptian inventions, to the importance of medieval Muslim intellectuals in transmitting ancient thought to the West and helping Europe emerge from the Dark Ages. This has fed a sense of injustice about who is credited for historical innovations, in some cases driving people to believe that the reality of Atlantis has also been covered up. In this perspective, the denial that Atlantis was an actual historical place is part of the same white supremacist worldview that refuses to credit non-white peoples for their contributions to culture.

Plato's use of Egypt in the allegory lends fuel to this theory. What if Atlantis's 'history' had been preserved in the East, but suppressed in the West, where it was dismissed as a 'mere' allegory invented by a Western philosopher? This would echo what happened in Plato's story itself: Atlantis, forgotten in Greece, was remembered by religious intellectuals in Egypt. The events of actual history have added an extra layer of resonance to this idea. Much ancient Greek history and philosophy were lost to medieval Europe, and were only rediscovered there in the later Middle Ages because they had been preserved and studied in the Islamic world. The idea that Atlantis's true nature goes unacknowledged in the West therefore chimes with both Plato's story and the events of subsequent history. Yet those who are inspired by this to believe Atlantis was a real ancient empire are, unbeknown to themselves, re-enacting the role of Plato's invented priest at Sais. In their efforts to rescue the story from cynical academics who declare it to be fiction, they end up playing characters in a modern version of that story.

★

Of all those who have helped to build the picture we see in our mind when we hear the word 'Atlantis', though, none has contributed more than Donnelly's famous contemporary, Jules Verne. In Verne's novel *Twenty Thousand Leagues Under the Sea*, Captain Nemo and the narrator of the book, Professor Pierre Aronnax, exit their marvellous submarine the *Nautilus* in standard diving dress – metal helmet with glass porthole, boots shod with lead – at a depth of 150 fathoms. Here they enter an underwater realm of forests and mountainous peaks, until they happen upon a long-lost underwater city which stands in ruins. Aronnax describes how

> There indeed under my eyes, ruined, destroyed, lay a town – its roofs open to the sky, its temples fallen, its arches dislocated, its columns lying on the ground, from which one would still recognise the massive character of Tuscan architecture. Further on, some remains of a gigantic aqueduct; here the high base of an Acropolis, with the floating outline of a Parthenon . . . Where was I? Where was I? I must know at any cost. I tried to speak but Captain Nemo stopped me by a gesture, and, picking up a piece of chalk-stone, advanced to a rock of black basalt, and traced the one word: ATLANTIS.
>
> What a light shot through my mind! Atlantis! the Atlantis of Plato, that continent denied by Origen and Humboldt, who placed its disappearance amongst the legendary tales. I had it there now before my eyes, bearing upon it the unexceptionable testimony of its catastrophe. The region thus engulfed was beyond Europe, Asia, and Libya, beyond the columns of Heracles, where those powerful people, the Atlantides, lived, against whom the first wars of ancient Greeks were waged.

It is impossible to overstate the importance of this passage in defining the 'Atlantis aesthetic' that has prevailed up to the present day. The

Disney film *Atlantis: The Lost Empire*, which is many children's introduction to the idea of Atlantis, borrows its set-up from Verne. A submariner journeys down to the deep seabed, and finds the faded glory of a forgotten civilisation.

Verne picked and chose the details in which he wanted to follow Plato – the location of Atlantis, and the fact it fought the Greeks in their earliest wars – and those in which he wanted to diverge. He departed from Plato in describing the well-preserved visible ruins of Atlantis: 'a perfect Pompeii beneath the waters'. It suited the drama of his story. Plato needed Atlantis to have disappeared entirely. Verne preferred it still standing.

Verne also describes classical architecture, with some very Athenian-sounding features: 'an Acropolis', 'a Parthenon'. There is no aesthetic distinction between Athens and Atlantis. Instead of having the architectural style of a mysterious 11,000-year-old empire, Atlantis looks just like the place its creator, Plato, lived in. Like every picture of Atlantis that resembles classical Greece, it seems to be accidentally admitting that is where the 'lost city' was really born.

Twenty Thousand Leagues Under the Sea was only a century and a half old. Was the provenance of my childhood dreams really as shallow as that? Had I been had by Jules Verne? Was the Atlantis I longed for a con?

★

The English poet and novelist Robert Graves believed the island of Pharos might be the location of a historical Atlantis. Seeming to believe in the internal logic of Plato's story, Graves referred to 'the Egyptian legend of Atlantis' which 'seems to date from the third millennium BC'. He connected Atlantis to the Keftiu or 'sea-peoples' who supposedly built the harbour walls of Pharos, and saw Plato's story as deeply linked with the 'flood legends' of Mesopotamia from which the biblical story of Noah had arisen. He also said that 'Several details in Plato's account, such as the pillar sacrifice of bulls and the hot-and-cold water systems in Atlas's palace, make it certain that the Cretans are being described, and no other nation.'[3]

Graves's certainty that the Atlanteans were really the Cretans runs into the same problem as every other literal interpretation of Plato: he bases it on a small number of shared features of the two civilisations, while ignoring the dozens of areas where there is no overlap whatsoever. This includes, of course, the location of Atlantis, which Plato specifies was 2,000 miles or more to the west of Crete, 'beyond the Pillars of Heracles'.

Plato's phrase 'beyond the Pillars of Heracles' is typically interpreted as a geographical reference to an area that lies past a certain landmark, or rather, a sea-mark: a threshold between water and water. In ancient Greek writing, the 'Pillars of Heracles' usually referred to two high promontories either side of the Strait of Gibraltar. According to myth, the 'Pillars' were built by Heracles himself to commemorate his tenth labour, the capture of the monster Geryon's cattle. The northern pillar was the Rock of Gibraltar, and the southern pillar either Monte Hacho (near the Spanish port city of Ceuta on the African coast), or Jebel Musa, just over its border in modern Morocco. Some have argued that Plato could have meant a different sea-channel. It could have been a very narrow one, like the Strait of Messina which divides Italy from Sicily, and was the probable inspiration for the rocks of Scylla and Charybdis; or a wider one, the Strait of Sicily, which separates the island's westernmost tip from the coast of Tunisia. Perhaps it was the Strait of Otranto (between the heel of Italy and what is now the Albanian coast), which marks the entrance of the Adriatic. If we interpret the Atlantis story as history, then the only one of these that could possibly make any sense is the first one, the Strait of Gibraltar, because of Plato's assertion that Atlantis was 'larger than Africa [i.e. North Africa] and Asia [i.e. modern Turkey] combined'. The Atlantic is the only place an island that large could fit.

But what if physical geography was the last thing on Plato's mind? Reading ancient texts is a perilous undertaking. Much is lost on us. A phrase might be beguiling, containing winks and nods to meanings that are opaque to a modern reader. It turns out that this is probably the case when it comes to the phrase 'beyond the Pillars of Heracles'. As much as the Strait of Gibraltar was a well-known physical place, the Pillars of Heracles also had a powerful underlying meaning: the boundary between the known world – the finite world, known to the Greeks – and the world beyond it, which was shrouded in a fog of unknowing, and was potentially infinite, and therefore frightening. To dive into water is to transgress such a boundary, into a world you do not know, and in which you do not belong. Describing a place as *beyond the Pillars of Heracles* is to locate it in an unknowable realm: a space where we dare not venture, or, at least, have never been. What if Plato's Pillars of Heracles is a philosophical reference to a different, mysterious domain: a realm of fiction? 'Think of the Greeks,' says the mathematician Robert Kaplan, 'as sailors, like Odysseus, but coast-hugging sailors, in the Mediterranean, not going through the Gates of Heracles toward the infinite beyond . . . They were puzzled by the infinite, out there. They were troubled by the idea of two lines which never met.'[4]

Two lines which never met. I was beginning to see such pairs of lines everywhere, like the twin bands of sky and sea kept apart by the horizon. There was the thought of Atlantis as myth, a delicate thread from a mind of the past, and running alongside it was the notion of Atlantis as a real place. There was my father's lifeline, and my own, the line which has lost its parallel and now runs beside a ghostline to an unknown end. There were the lines we write in praise of the sea and the tripwires by which we destroy it.

*

In his *Encyclopedia of Dubious Archaeology*, the academic and archaeologist Kenneth Feder lists the many physical attributes of Atlantis as described by Plato. There are fifty-three of them in total, most of them found in the *Critias* – and they range from the colour of the stones to the gigantic size of the island, its hot and cold springs, its system of bridges and canals, and the materials from which its buildings and walls were made. When somebody claims to have found 'the real Atlantis', as Feder observes, they tend to cherry-pick a small number of descriptors from this long list, and then proclaim that the similarity between their site and Plato's Atlantis is undeniable or even uncanny. It is rare, he says, for an Atlantis-seeker to take a systematic approach.

In line with this tendency, the popular author Graham Hancock has written that 'there are four essential ingredients in Plato's story'.[5] Not only is four a seemingly arbitrary number of features to arrive at, but some of these 'ingredients' are not even found in Plato's work. Hancock says the destruction of Atlantis was 'the result of a global cataclysm', which Plato did not say. This would have defeated the whole point of the story, which was that Atlantis was destroyed because of its hubris. A global cataclysm would have also devastated Egypt, which plays the role of preserving the memory of Atlantis's fate.

This also presents a particular problem for those who, like Hancock, claim that we should not dismiss Plato's tale as a 'mere' allegory, because myths and legends can preserve cultural memories of past events. Yet the Atlantis tale is not a myth or legend. There is no evidence of the story, or even of the word 'Atlantis' in the sense of an island, having existed prior to Plato writing his dialogues. What we have is a detailed account, split across two works, that is traceable to the hand of a single person. A myth would be expected to have lost much of the material detail, but preserved a vague echo to map on to some specific historical event – yet Plato's story is intricately detailed. Only the secondary sense of 'myth' is applicable here: that of a widely held but non-historical belief. Artistic fiction partly inspired by contemporary events is not the same as the survival of folk memories.

None of this has prevented modern people from searching for the remains of a physical Atlantis. In 1931 the Woods Hole Oceanographic Institution of the United States announced, to some local media fanfare,[6] that they were about to send a team on a voyage to survey the Atlantic seabed – an unspecified area potentially tens of millions of square miles in size – in the hope of finding signs of the lost empire. The mission was aborted without success, but similar ones have periodically continued to attract support.

In 2005 the Iranian-American writer Robert Sarmast convinced the film company TMC Entertainment to spend tens of thousands of dollars documenting his quest to find the acropolis of Atlantis seven miles off the coast of Cyprus, at the unlikely depth of 5,000 feet.[7]

'We have definitely found it,' Sarmast told journalists at Reuters.[8] He definitely hadn't.

In 2017 the film-maker and journalist Simcha Jacobovici made a documentary about his quest to solve the mystery of Atlantis. The film begins with an awkward sequence in which Jacobovici goes to talk to his friend, the Hollywood director James Cameron, about the project. Cameron, a seasoned underwater explorer and technological pioneer and one of the few people to have visited the deepest parts of the ocean, makes it clear that he believes Plato's story is best read as a cryptic allegory, but wishes Jacobovici all the best with his searches.[9]

These three quests all have something in common. They are nautical adventures: contemporary tales of men setting out on ships, each believing that they might at last be the one to solve the 2,400-year-old Atlantis 'mystery'. They are stories of mariners' hubris, and in this they echo the expeditions of the doomed Atlanteans which inspired them.

Hermits

Widewater Lagoon, England

I realised I'd lost everybody. It was crazy how fast it had happened. I looked around through the jade murkiness of the sea and tried to spy a trace of anyone's fin. There was nothing. The dancing seaweed I could see within the short range of my vision began to look sinister. I knew this combination of image and feeling: I had seen it in a childhood dream, but now it was real. There was no waking up. I was getting my first pang of the old fear: the fear of entanglement on the seabed. It was the fear of Rán.

It had been weeks since I'd got certified as a diver and I was itching to get back in the water. I'd joined the local diving club, Ocean View. It was high summer. I presumed I'd be able to tag along on a dive straight away.

It wasn't that simple. A number of factors need to align for a dive to take place in the sea. These vary from place to place, but they must always be taken seriously. On our stretch of coast it is only safe to dive at slack water, a particular moment in the tidal cycle which, off our beach, begins about four hours before high tide. We had to wait for this to happen during the day, and at a time when people weren't at work. In addition to this, the wind needed to be very light, or ideally absent, which didn't happen often in this part of the world: it is typically buffeted by a prevailing wind from the south-west. And the wind direction was important. Even a gentle breeze coming from the east was likely to wipe out the visibility. In the Channel, easterly weather brings silt from the estuaries of the big river mouths that lie in that direction, whereas winds from the west tend to bring in the clearer water from the Atlantic, and with it, better conditions.

I sat at home surrounded by Dad's possessions, wondering whether it was doing me any good. In his wake I was uneasy. His things – and in particular, the things he had treasured, like his paintings and record collection – seemed to be imbued with a trace of his life, which, now he was gone, felt like a tang of death as well. The place was a shell he'd inhabited. I wanted to be in the sea, with its inhuman furnishings and its lack of enclosing walls. Covid-19 lockdowns intensified the claustrophobia. I was desperate to get out diving. *I'm turning into a hermit*, I thought.

I had almost given up hope when I finally got the summons I'd waited so long for: a message from Teresa at the dive club. Cold with excitement, I read:

> *Hi Damian. The weather forecast looks good so a few of us are going to meet at 9 a.m. tomorrow for a dive at Widewater Lagoon. Just letting you know in case you wanted to come along and join. We'll meet in the car park and if the conditions are good, we'll dive. Maybe see you tomorrow. Teresa*

'If the conditions are good, we'll dive.' It was such a simple sentence: spare, short, unremarkable. But what it meant to me was huge and complex. Teresa had only sent me this message because I was now a qualified diver. It meant I had reached the first peak on the unlikely journey I had plotted for myself.

I quickly replied that I'd be there, then drove to the dive shop – struggling not to exceed the speed limit – to hire a tank full of air. The good news was that the water was likely to be in the twenties, its peak temperature for the year. It meant I wouldn't have to hire a drysuit but could use the wetsuit I'd bought instead. I would have freer limbs, and fewer buttons to press: less on my mind.

When I pulled into the car park the next morning, the other divers were already there. Mark had grown up nearby and was about my age. The other three – Teresa, who was originally from the West Midlands; Kenny, a Scot; and Paul, a local like me and Mark – were my seniors. But the only 'age' that really mattered out here was our sea-age, the amount of time we'd spent in the water. By this reckoning I was not just the youngest, I was an infant who still needed babysitting. It was kind of them to let me come out with their group.

Several things unified these divers. They had a shared aura. They looked like a squad: a group of friends who were ready to set about a complex task together. All four of them had the same bearing. They were excited but serious with it. They were all wearing what appeared to be black tracksuits, which I then realised were their diving thermals – tops and trousers made from specially engineered fabrics that stay warm even when waterlogged. These are a crucial bit of kit because if your suit leaks and you are far from the surface, or need to make decompression or safety stops, it could be disastrous. You cannot afford to risk getting cold in your suit.

We started to put on our gear. I was jittery, not having dived in open water since I'd passed my basic tests. Phil, who I'd learnt was Teresa's husband, came over to give me a hand. He wasn't wearing one of the 'tracksuits'.

'Not diving today, then?' I asked.

'Oh, no, I don't dive.'

'Oh, right.'

'You might want to attach your inflator hose,' said Phil.

'Ah, bloody hell! Cheers,' I said.

Given the way he'd instantly spotted a problem with my kit, I reasoned that by 'I don't dive' Phil meant he was an experienced diver who had stopped diving for some reason. I'd thanked him for helping me, but inside I was annoyed that I needed to have something so basic pointed out. I was learning to welcome assistance from other divers, but slowly. There was so much to remember that you had to be humble and accept a helping hand, whether it was someone offering to hold your cylinder up while you put on the BCD, or mentioning that you'd forgotten something. In my imaginings of diving, before I'd started training, I'd always thought I'd quickly learn to do everything perfectly, and by myself. That wasn't a helpful perspective, but it was one I was having problems shaking off. Diving, at least at the basic, amateur level, involves a social contract of sorts. You help each other, and try to have the good grace to smile while you're doing it.

We trudged off down to the shore and stopped so Phil could take a picture of us. Then we walked into the sea, spat into our masks and rinsed them in the water to prevent them fogging up. We pulled our fins on in the shallows, and started to swim out towards the buoy where we would drop down and take a look. Phil waited on the shore, looking through a pair of binoculars. He would keep an eye on our surface marker buoy and, if we disappeared or were gone too long, he'd call for help. I looked back at Phil a few times as we swam off. His focus on us was total. *See, he respects the sea*, said Dad.

At the buoy we dropped and headed south underwater as the current did its best to push us westward. This was an early lesson for me in the deceptive qualities of the sea: a gentle breeze had been blowing out of the south-east, causing wavelets on the surface which made it look as if the water was going in that direction. On the bottom it was a different story, with the current nudging us the opposite way. The relationship between the wind and the current is not straightforward. They are neighbours, but they each have wills of their own.

I did my best to control my movement by finning alone, but occasionally I grabbed at the rocks to steady myself. I wasn't happy about it. Every time I have grabbed at a chunk of the reef in order to hold myself in place against the current, I have thought about the damage I cause.

Great as it was to be back in the water, the visibility wasn't good: perhaps ten or fifteen feet. We headed south and the sea around us began to grow more and more opaque. Looking ahead didn't reveal much. There was more to be seen on the seabed, which was less than four feet beneath us. I spotted a large orange starfish, *Asterias rubens*, also known as the sugar starfish or common sea star. They can also be purple but this one was the more typical grapefruit colour. I stared at it to see if I could perceive it moving slowly. It was in no hurry. Tilting my head, I could just make out some of the tiny, pale tube feet that cover the starfish's underside. This orange animal the size of a hand has thousands of hands of its own. The tube feet operate in concert, using a combination of hydraulic power and a glue-like secretion from every foot to move in the desired direction. Desired, but not overthought: the starfish has no brain, and finds its way using extra-sensitive feet on the ends of each arm, each of which is accompanied by a small light-sensitive 'eye-spot'. And of course, it has a powerful appetite. Drifting over them on the current, it is hard to believe that their underside is engaged in a constant Mexican wave of brainless, but potent, intent.

Starfish are senior earthlings. The first ones arose around 450 million years ago in the Ordovician period, with a common ancestor in a starfish-like creature, *Cantabrigiaster fezouataensis*,[1] of which a fossil was discovered in southern Morocco that dates to around 480 million years ago. These beings have been around a long time and have not needed to move fast in order to achieve it. The common starfish's typical prey are bivalves: animals less mobile than themselves. Starfish are slow but relentless hunters and they are good at what they do. They use their tube feet to pry open the shells of molluscs like clams and mussels, and they will exploit even the tiniest gap in their victim's defences. Starfish can extrude their stomachs through a shell opening less than a twentieth of an inch wide, through which they dribble enzymes which digest the mollusc's flesh. Hunting in this way, and able to go for up to six months at a time between meals, a single common starfish might live for as long as eight years. One female starfish can produce several million eggs.

Looking up, I realised I had lost sight of my buddy, Teresa, and the other divers. I had stared at the starfish too long. I scanned about for them and checked my watch. A minute had passed, so I did what the protocol demanded and surfaced to wait for Teresa to join me.

After a short while at the surface I wondered whether I might be able to see the others, so I let a little air out of my BCD and dropped my head below to take a look. There was no sign of them. I returned to the surface and waited. A minute later they emerged.

'What happened?' asked Teresa.

'Oh, sorry. I was looking at stuff and I lost you guys.'

'OK. Best stay close together. The viz isn't great.' It was a gentle rebuke but I took it on board and felt bad for losing my buddy.

'OK, yep,' I said.

'Shall we drop back down and see how it goes? Can't see us staying down there for too long today,' said Paul.

We descended again and the conditions continued to get worse. After another five minutes Teresa and Paul decided to end the dive and gave the thumbs-up signal.

We surfaced and swam ashore on our backs. Phil helped us out, offering to carry our fins or weight belts if anyone was tired. He walked alongside me, ready to lend a hand. I noticed he had a slightly stern look on his face.

'When you came up, you went back below,' he said.

'Ah. Yes I did.'

'Why did you do that?'

'Well . . .' I realised I didn't have a good reason. 'I thought since we were only in shallow water, I could just take a look below and see if I could see anyone.'

'You don't do that. You come up, you wait for your buddy.'

'OK.'

'Otherwise, think about it. What happens if you go back below, and they surface, and then they can't see you? That could cause serious problems. I was watching and I thought, *What's he doing?* I'm thinking, *Hang on, you're a qualified diver, and you've failed to wait at the surface for your buddies to reunite.*'

I didn't have anything to say in reply. I felt like an idiot. It had seemed harmless at the time, but Phil was right. I had made a mistake. And on my first dive with these guys, who had been kind enough to take a novice out into the sea for a debut dive off his home shore.

'I'm sorry,' I said.

'Don't apologise to me,' said Phil. I'd heard almost these exact words before, from the divemaster at Wraysbury. 'The protocol is there for a reason. It's to keep you all safe. That's why you do what it says. You've made a mistake, you learn, you don't do it again. That's the only reason I'm mentioning it. No need to say sorry. It's all a learning curve.'

I drove my cylinder back to the shop, happy I now had a sea-dive under my belt but disappointed that it hadn't gone better. I didn't want the search for Atlantis to be defined by worries about my competence as a diver. I'd improve next time, I promised myself: stay on course and follow the protocol.

At least I'd got back in the water. It was good to have breathed alongside the starfish again.

*

A few days later we dived at the same spot. It was to be a lesson in how no two dives, even in an identical location, are ever alike. The water was transformed. The current was gentler. We could see twenty-five or thirty feet ahead, so keeping track of each other was easier. It was so unlike the previous dive that I had trouble believing we were in the same place. It was like the difference between standing in a field on a summer's day compared to in winter fog, but this change had occurred so rapidly. The moods of the sea swing fast.

Having got our bearings, we hovered slowly across sparse prairies of thongweed. I was relieved to see that in spite of the recent storms, so much of it was still attached to the seabed, swaying with its golden colour that seemed to turn to burgundy when in shade. In the sashaying current the stringy weeds looked curiously feeler-like – alive in an animal way – and the odd strand tickled the exposed part of my top lip, just under the mask. Sometimes it is difficult to maintain, in your mind, a line separating the plant kingdom from the animals. Animated by the water and by its yearnings for light and nutrition, the seaweed dances in a rhythm that communicates an inner life. And often, like a tree with its insects and birds, it is itself home to abundant lives: tiny hydroids and vegetarian invertebrates. I thought of these seaweeds lying on their sides at low tide – horizontal – fallen asleep. It brought back my old childhood thought of how, for sea-life in the intertidal zone, dryness is a sort of equivalent to darkness: a second kind of night, defined by want of moisture instead of light. The absence of the substance that makes its 'day'. The turn of the tide is a dawning of water.

Squadrons of hermit crabs were scuttling in a hurry towards the east, every one of them perfectly fixed on the same line of bearing. Their formation was loose, making the name we have given them – the word for hermit crabs means something like 'hermit' or 'loner' in every language in which I have looked it up – just about excusable. But they had some secret common knowledge, each one driven on by a prevailing purpose that was a mystery to me. Why should I understand? The hermit crabs have lived on Earth for 200 million years. Their earliest ancestors used the shells of ammonites, whose petrified bodies stretch back in the fossil record to the time of giant armoured fish, when Antarctica and Australia were close neighbours whose shivering and simmering fates had not yet diverged. What staggering lineages. When we named the hermit crabs after ourselves, we were naming our elders. The insolence of it. These tiny, frugal and timorous fiddlers hail from a stock that makes humans seem like newborns. Their artfulness has seen them through the age of flying dinosaurs to the time of apes who shoot themselves at the moon. And as someone who once lived in the back of a second-hand van, I also feel an affinity with them as mobile hermits in their pre-owned shells.

I peeked under a large flat rock and was greeted by the telltale crimson eyes and squat blue face of a velvet crab, sticking his dukes up at me, a martial posture, ready to fight or die. I am always astounded by the fearlessness of these animals, valiant in the face of an alien creature hundreds of times their size, holding a mesmerising light in its hand, capable of their seizure and dismemberment. An ogre at the gate of their home. We will never know what this is like: there is no animal that can tower above a human as we loom over little things. The crab knows the odds and steels itself to fight. It's clear that the torchlight fixes its mind, but the specifics of what it feels are lost to me: a page in the endless secrecy of the crab. Let me never recall the crab without remembering its courage. People mock moths or squid who are drawn to their deaths because they cannot resist the siren pull of a torch, when humans spend their lives stupefied by screens and TVs. Each night, when insects are attracted to our phones, we are reminded of the brash youth of our technology: these animals are mistaking our trinkets for the moon, the only thing that glowed this bright through the prehistoric night. Fools' moons, our torches are.

Entranced by the crab, I had forgotten to mind the progress of the dive. I looked around to find I had been left behind by the others. A jolt of panic was followed by flustered kicks as I tightened my core and caught them up, upset that I would probably never see the crab again. Scuba diving involves such a delicate waltz of attention between the life of the sea and your band of buddies. Being interested in things is an occupational hazard.

Nearby, a giant whelk was on the move. Its exposed skin was Dalmatian-esque – a beautiful smooth white marked with charcoal strokes, a dappled finish worthy of a marble tabletop. I was used to seeing these animals cooked, shrivelled, a sad parody of themselves. They were a favourite snack of my great-grandfather. On the kitchen worktop there would be a bowl of them, puckered into chewy little chestnuts of meat, their sleekness obliterated by boiling. I'd had no idea their skin was so beautiful in life.

There were small grey shannies and juvenile sea bass everywhere, darting about and occasionally curling in fear. Then I saw, with its outsized fighting claw, a hermit crab of the species *Diogenes pugilator* – meaning 'boxer', because of its hefty swinging left 'hand'. This club-like appendage is supposed to give the crab an advantage in contests fought against fellow crabs over shells. It was given the other part of its name, *Diogenes*, in honour of the philosopher Diogenes the Cynic, who lived in a barrel in the Greek city of Corinth.

The word originally used for Diogenes's barrel was *pithos*, meaning a large storage vessel, an enormous clay jar. It might have held olive oil, wine or grain, before finally becoming the home of a topless philosopher. What it had not been was the sheath of another living thing. Diogenes would have borne a closer resemblance to the crab if he'd gone about in the pelt of an elk or bear. The hermit crab doesn't sit underneath a roof of moulded clay. It peers out from beneath a hard cloak left behind by the dead. It has got dead man's shoes.

The hermit crab's shell is a project on which many beings have collaborated, and not all at the same time. It is often furry with a moss-like living fluff and crested with anemones, sometimes even playing host to a commensal ragworm that lives within the shell alongside the crab. A sea-snail originally made – or rather, partly was – the shell, which has been re-embraced beyond its original purpose, and a certain 'snail-ness' is required by the new inhabitant. There are at least 800 species of hermit crab, most of which have an asymmetric abdomen which can 'snail' its way backwards into the shell.

It is a weird sort of homage to an absent, dead predecessor, and I find it troubling: am I, I suddenly wonder, getting more like my dead dad because I have been living in his house? It is an unsettling notion, like being retrofitted to a space abandoned by a previous life. Plato's idea of Athens is of a state that is somehow 'plugged in' to the space left behind by the more perfect, primeval version. Are we all hermit crabs, shuffling backwards into a place not of our own making, concealing our vulnerable parts beneath the cast-off armours of former cultures?

The 'fur' mat that covers the hermit crab's shell is actually a species of marine hydroid called *Hydractinia echinata*, also known as snail fur or the hermit crab hydroid. It may look mossy, but hydroids are tiny predatory animals, so this fuzz is really a carpet of hungry animal life. It consists of three types of polyps: club-shaped eating polyps, reproductive polyps called gonozooids, and defensive stinging polyps called dactylozooids which, in spite of their name (suggesting a resemblance to fingers or toes), are long and thread-like: stringy but dangerous, like the tentacles of a jellyfish. The larvae of the hermit crab hydroid are minuscule crawling animals which can sense that a moving gastropod shell is nearby, causing them to clamber aboard and begin shapeshifting into the multifaceted adult version. In places, the fuzz of *Hydractinia* seems to cover everything. It gives an aura-like hazy glow to the shells and rocks on which it lives, coating their calloused surfaces with a softer living layer. At times, as columns of gold sunlight scan over it in the water, it has the look of a dense layer of short hair, like the hair of a mammal.

The shell of a hermit crab becomes a 'natural ruin' – a structure recolonised by nature to create a new, hybrid form. But this ruin is lived in, as well as on: a little houseful of life under the sea. There weren't any human ruins at Widewater, beyond the odd nubs of wooden groynes that occasionally transected the shallows. But I had found ruins of a different kind.

I had set out to look for a human Atlantis, but the more I looked around the more I saw examples of inhuman ones – camouflaged, animal – and I felt a new pang: a risk of losing interest in my original obsession with underwater human structures. With every dive, every stride across the rock pools, the fixation with what we humans had left in the sea felt a little more short-sighted. It would be absurd to ignore the life that was everywhere.

Better to learn to pay attention to all of the layers at once, I thought. Why shouldn't we be able to appreciate, simultaneously, the old and the new, the human and inhuman, the original thing and that which lives upon it, and all these things together? The Atlantis motifs had supplied a romantic framework from the human side, an allure that had sent me diving in the first place. I felt the thrill of these powerful influences colliding, merging into a new whole which I hadn't foreseen or intended. Plato and Jacques Cousteau had collaborated accidentally on the experience I was having. As the ruin and the plants that grow upon it become of a piece, as the crab shell and the hydroids form a single mobile community, I too was a product of many forces, propelled through the water by an old Greek story as much as I was by my cylinder and my fins.

Not every species of hermit crab uses a hard shell for protection; some choose a living sponge. The crab initially lives in a shell, which a young sponge settles on and slowly eats away. Unlike shells, the sponge 'grows up' alongside the crab, removing the need to hunt for new, larger carapaces, a perilous pursuit during which the soft abdomen is exposed. But sponges can attract more predators and grow to enormous sizes, leaving a relatively tiny crustacean encumbered with an immense mobile home.[2]

You shouldn't keep the possessions of dead people, the Romany culture warns us. I had inherited this moral from my family, but faced with the reality of having to jettison the belongings that had been precious to my dad, I struggled to do as it commanded. It was a cold principle, demanding contempt for the invisible trace of the dead on the things they had left behind. How do you not keep dead people's things when everything on Earth, one way or another, has been touched or owned by or channelled or contemplated by something dead? How do you keep anything? How do you not keep everything? Unlike ancient nomads who had no space for inherited hoards, or sea-creatures forced by the communal nature of the ocean to travel light, I had the luxury of wondering what to do.

As we got out of our equipment, it was once again Phil who came over to help me. I was learning to appreciate the assistance. A dive takes a lot out of you. By the time we had donned our gear, walked down the beach, swum out on the surface, dived, swum back and walked back up the beach, we had been in our kit for as long as two hours; I felt spent, although I was trying to put on a brave face.

'Was there a reason you stopped diving?' I asked Phil.

'I never started!'

'What?' I was shocked. 'You're not a diver?' I asked Phil.

'Nope! No way I'm bloody well getting in there. You lot must be mad!' Phil chuckled as he helped lift my cylinder off my back.

I was stunned to learn that Phil wasn't a diver. The fact he was husband to Teresa, who was so accomplished in the water, and that he had an eagle eye for the equipment and processes of diving, had led me to generate a backstory that he'd given up diving after a long career because of some health worry, or because he'd done it all and simply had enough. It turned out Phil had been an engineer before his retirement: to him, scuba equipment was all built from concepts he understood very well. It made me think about saturation diving, where the people controlling the diver's 'umbilical' – the cable which supplies their vital needs of air, warmth and communication – are just as fundamental to the process as the divers themselves. No diver is an island.

When I got home, the place felt different. Something silent, but huge, had changed. I looked at the walls Dad had painted, his artworks that hung on them, his record collection. *A shell, a living fuzz, the sound of the shell* . . . Was it an absurd analogy? No, it was a healing one. I was the new life in the old shell. The hermit crabs had informed me that, maybe, it was all right.

The Lessons of Stone

Lake Van, Turkey

There are places on Earth that were formed before there was oxygen in the air, and in them were things that lived.

The Pilbara region of north-western Australia is one of them: it has been called the oldest place on the planet. In 2013 the geobiologist Nora Noffke, head of a team of excavators who had been working in the Pilbara, announced the discovery of fossilised bacteria dating to perhaps a billion years after the formation of the Earth, when our world was only a fifth of its present age. These microbes fed on elemental sulphur, turning it into hydrogen sulphide, a key ingredient in the formation of the environment and the emergence of later and more complex forms of life. They multiplied. Over the cycles of their lives the bacteria deposited adhesive compounds that bound their colonies to the stone, eventually creating 'microbial mats' – a kind of living rock. These are the fossils the scientists discovered. You can go to this place and touch them, place hands on something that thrived in a time when you could not even have breathed.

These microbial mats of bacteria were a new thing on the land: a hubbub of tiny lives, a thrumming 'city' built upon the ruins of the old order. Each one was a living tel, a living palimpsest. The oldest of them are 3.5 billion years old. To put that number in perspective, the span of time separating us from Plato – about 2,400 years – would need to be repeated almost 1.5 million times. The fossils of these stromatolites are warning us about something. *There wasn't always oxygen, you know. Be careful what you take for granted.*

Mind-boggling as the antiquity of these fossils might be, the first ocean was older still. The earliest seas are thought to have accumulated as many as 3.8 billion years ago, and they were likely rich in oxygen long before it built up in the atmosphere.[1]

Fast-forward to our time, and around half the oxygen we breathe is produced by the sea. This idea has not taken root in public consciousness with the same solidity as rainforests being the 'lungs of the Earth'. Perhaps it is easier for us to think of forests as giant lungs, their trees resembling the alveoli within our own pulmonary system, their broccoli-like texture an absorbent inner skin. On the other hand, because water drowns our own lungs, perhaps the notion of the sea producing the breath of life seems somehow ridiculous. The sea suffocates us: how can it also help us breathe?

★

Plato wrote his tale of Atlantis more than two millennia ago, but it was far from the first of its kind, and not just in the sense that humans had been telling stories of warring empires long before his time. Not all accounts are written by people. There are other kinds of records, left by other kinds of lives.

Across vast spans of time, inhuman forces on planet Earth have been authoring their own stories of the rise and fall of lives and powers and waters. They sculpted friezes for us to decipher, and left them cast in stone; instead of using alphabets they hieroglyphed with their bodies. In place of pictograms they embedded their skeletons in rock. As cryptographers and linguists have figured out the scripts of Sumer and of Mycenaean Greece, so palaeobiologists can 'read' these mysterious signs. The once-living rocks of the Pilbara tell a story of the land, and there are others like them.

In 1990–91 submariners discovered a series of structures on the bed of Lake Van in eastern Turkey. They looked like a cross between haggard desert ruins and the wind-sliced pinnacles of Death Valley.

Some of them were up to 130 feet tall – taller, perhaps, than the ziggurats of Babylon, the possible inspiration for the biblical Tower of Babel.

The 'towers' in Lake Van are called microbialites, a word that means 'microbe-stones' or, more poetically, 'the rocks of the little lives'. They are solid forms on the lake bed formed by coccoid cyanobacteria, a group of micro-organisms which work to create structures out of rock and sedimentary deposits. We might deem them to be among the first living builders on planet Earth. The microbialites formed by these cyanobacteria look like ruins; but whereas a ruin is erected by humans, then damaged, and perhaps finally reinhabited by other life, a microbialite is built by non-human life, which dwells in it from the start. Microbialites are living rocks, so the dead microbialite is, perhaps, more like a skeleton than a ruin. In some sense it resembles a coral reef.

On the shores of Lake Van there are large, dead microbialites which used to be in the water, and which now jut up from dry land due to the slow shrinking back of the lake. They stand baking in the heat, tiered structures full of columns with clearly discernible storeys – like the floors of a building, built for living in. Turrets stand atop them. Near their feet are little caves with columns of rock between their floors and their pillars. Glance at them quickly and you would be forgiven for thinking these were the work of human hands and tools. They look excavated. They seem deliberately formed. These ghostly edifices, risen from the lake into the sun, are a skeletal Atlantis.

There is something frightening about the spindly, organic pillars of the microbialites. Aesthetically they lie somewhere between the work of humans and spiders, yet we know they were built by something older and more primitive than either: by a thing of great patience, working away without a thought of what we would some day think of its projects. Some of these microbialites are sixty-five feet tall and up to 16,000 years old. There are photos online of men clambering about inside them as if exploring a temple. One picture, taken by Ali Ihsan Öztürk, shows a man in a baseball cap and backpack, standing on the 'roof' of a huge, tiered complex. He stares down nonchalantly at his phone. Below him, another man dressed all in black is picking his way along the strata of sediment beneath the chambered structures. They were made in the water; the water has receded and left them dead; the living walk about on the dead remains; and the water will return, perhaps in a time when we are dead and the microbialites are reinvested with a new ecosystem, giving them another life. In these still images can be read the cyclical nature of birth, destruction and rebirth in our world.

Human antiquity is nothing compared to the geological scale: the depth of 'deep time', the time of the rocks and fossils. But the age of our own history strikes us more profoundly, because it is easier to contemplate. A span of a hundred lifetimes makes more sense to the human mind than a million generations of jellyfish or the 4 billion years that have passed unseen within a rock of the Acasta Gneiss of northern Canada. The word 'antiquity' conjures statues and skeletons that have waited out the ages of mortals and our dusty recorded history. The Sumerian, Akkadian and Babylonian tales are thickly covered in this patina of age. They come from Mesopotamia, a Greek name for a non-Greek place, meaning 'between-the-rivers-land', most of which lies in what is nowadays Iraq. In the oldest of these stories, there is a recurring theme: that of a great and devastating flood.

Perhaps the most famous flood survivor is Noah, he who built the ark into which the animals went two by two. Noah, though, was simply one name — the Hebrew name — of a figure with roots far older than the Bible. Behind Noah lies a shadowy but identifiable character who has merged with several heroes over millennia. His existence testifies to the human obsession with floods. He appears as though gazed at through water, indistinct, called different names by everyone who catches sight of him. In Akkadian, he was Ut-napishtim or Uta-na'ishtim, meaning 'he found life'; in Sumerian, he was Ziusudra; and in Old Babylonian, his name was Atrahasis, which meant 'extra-wise'. The Greek name Prometheus is perhaps a rough translation of Atrahasis, and the name Noah itself might — just possibly — be a contraction of (Ut-)na'ish(tim).[2] What is certain is that clay tablets containing the Atrahasis epic, written in Old Babylonian, have been dated to around 1700 BC, and that their author's name was Ipiq-Aya. 'Rarely is an author named for such an early text,' the linguist and archaeologist Stephanie Dalley has pointed out.[3]

These stories are tough. They live in stone. Impressed into clay tablets in cuneiform ('wedge-shaped') script, they have survived through thousands of years of storms, earthquakes and human interference. Sometimes they have survived not in spite of destruction, but because of it. When besieged cities burned, the tablets were baked hard by the flames, in a crucible of war, making the physical text almost as indestructible as the tales it conveyed. Yet still, over time, there was damage done: stone was chipped, tops and bottoms and sides broken off. The result is that, even in transliterated form, the text itself looked flood-damaged: chunks of it missing, pieces of mosaic dislodged as though by water, riven by tides. The form comes to echo the content. Time has stolen the tale's completeness as water wears down all things.

★

The winter had come and gone. I was diving more often with the guys from the club. We planned to meet one Sunday in summer as a perfect morning rose up over our coast. The sunlight sighed off the eastern water, and in the breeze the sea was tranquil but alive. Its glassy wavelets moved in many directions, as if hurrying to convene. It was an image of the sea as a community of one, constantly converging on itself. A single cormorant drove a low, straight flight along the surface, until it met the sun's reflection and disappeared into the light.

We met up at eight o'clock. There were three of us – Kenny, Paul and I – smiling at the emptiness of the car park and the beach, the perks of coming out early. It took me a while to get suited up, as it always does after a break. My hands seemed ahead of my mind, piecing things together and reaching for the next bit of kit before I'd remembered I needed it.

As we got ready an old man with silvery hair ambled over to Paul and Kenny, leaning on his walking stick. He was wearing several poppies – little paper and plastic flowers sold every autumn in Britain to commemorate war dead and raise funds for living veterans.

'Be careful or you'll end up in the decompression chamber!' he said, and chuckled knowingly. Paul replied that 'the bends' was unlikely to be an issue as we weren't diving beyond twenty-five feet.

'Ah. I'm sure you'll be all right, then,' said the old man.

It turned out he had served in the Royal Air Force in his youth, and travelled widely. He had done some diving, many years ago, and studied the dangers. His mention of 'the bends' was less naive than I had presumed.

The old man wished us a good dive and went on his way, and we trudged off down the beach. After doing our checks we stepped into the water. I was briefly shocked when the cold brine flooded my gloves: according to my computer the water was thirteen centigrade, a temperature that always looks warmer on paper than it feels on the skin. I prayed none of it would creep under the seals of my drysuit. The surface was flat calm. We swam out a few hundred yards, got our bearings, agreed a three-part plan, and quietly slipped below.

The visibility was good and the current barely perceptible. Once I was neutrally buoyant it felt blissful to be back in the sea. I had another self now that only really existed there: a submarine me, my selkie side, a shadow that lived in the water. Again, it struck me that it was partly the matter of being forced to go for a while without language, bar the simplest of gestured signs. A pair of divers fluent in sign language would experience it very differently, but for me it is always startling to plunge into this way of being alive where things did not revolve around spoken words. It is almost akin to dreaming, but with a conscious autonomy over the body: a lucid reverie in another world. A way of being alive without everything revolving around your words. In that sense it is akin to being asleep, but with a conscious autonomy over the body: a lucid dream in another world.

I had come to love shallow-water diving. In the shallows, the joys of scuba – of being able to spend an hour or more breathing weightlessly under the surface, and of coming face to face with curious marine life – are combined with few of its stresses, because the surface is always there, a few fin-strokes above. Decompression isn't an issue. A three-gallon cylinder full of air seems to last forever, and if you can restrain yourself and keep reasonably still, the more reticent creatures start to emerge and investigate, and you notice the small things: tiny crustaceans scaling the fronds of weeds, perhaps a little cuttle that, were you swimming furiously, would almost certainly pass you by. When people poke fun at the shallow depths we've been diving in, it feels like they aren't just calling the diving shallow, they're calling me shallow as well. I am not a deep enough guy for them. Like these shanny-filled inshore waters, I am not risky enough, I am no challenge, nothing to brag about. It is fine by me. I agree with the diving organisations' dictum that a successful dive is one from which every diver returns unharmed. I do not want to risk my safety or the safety of my buddies. For me, the unacceptable risk is missing the sight of something beautiful because I was hurrying, or obsessing over my equipment, or doing deep dives for the sake of it, because the numbers in my log were more important than the life in the water.

On this dive there wasn't much 'big life': a phrase that refers to anything with a size between that of a small pet – a hamster, perhaps – and a human being. There is always plenty of 'small stuff': hermit crabs on their panicky errands, winkles and limpets on slow forays. Sometimes there are few of the larger beings around: the bass, the adult spider crabs, starfish and multicoloured wrasse, whose worlds we gatecrash. During such outings the absence of the large animals, with their dense gravitational pull on my attention, frees me to appreciate other things that often drift by underneath, uncontemplated.

I decided that today I would not miss the stones. Much of the seafloor here, in the places between the wide fields of sand and the reefs of bigger rocks, is a tapestry of little stones that have been rearranged by the sea ahead of every time we dive: a shifting mosaic which the English Channel never declares to be finished. It is a self-portrait of the sea: an arrangement, a work of the water.

Every stone is beautiful, an abstract sculpture. Igneous, metamorphic and sedimentary dwell alongside each other, born of fire, of great weight, of bones. They are of vastly different ages, elder and younger stones tumbled together. The youth of the chalk shines brightly beside the dark translucence of flint, in which tiny imperfections twinkle like far-off stars, and in a sense that is what they are: bursts of dust that exploded in a fit of physics long ago. When we look at them, as when we look at the night sky, we see the past alive in the present moment. It is a miracle. Split flints, with their inner resemblance to a picture of outer space or of deep water, speak of the depth of time from which they also arose.

On a long-time-lapse film we would witness the stones for what they are: ever-dwindling bullets of consistency in a whirling world of water and air, meteors burning up as they pass not through the atmosphere, but through duration. We would watch them diminish, perhaps from islands or mountains into pebbles, participants of shingle and finally sand, as the flicker of days and nights moved by so fast that they would register as nothing but one long half-light, an aeonic gloaming. We would see young rocks being born, the children of volcanoes. We would know ourselves as potential raw material for the chalks and shales of the future, and we would know to pay attention to more than the brief lives of animals, each one just a breath in our world's long dive through time.

The stones wait for me. Their patience is perfect; there is not a breath of restlessness in it. They will wait until both of us are unified in sand, and the peace of that future hangs between us like a wordless bond, a link that cannot be severed by bad luck, effort or war. Each piece of chalk, rounded into a miniature menhir of pallid stone, eye-holed with narrow caves, is a gravestone composed of the very lives it commemorates. It is a little monument to the unrecorded biographies of the sea.

I started diving in the hope of finding ruins, whether they be buildings or stories. I wanted clear arrangements of stone, or architecture lain out in a legend: the ghost-bones of worlds, even imaginary ones. I still wanted those things and was prepared to go below to find them, but there was something else as well. Now I could see that however they had fallen, whether or not they had been placed consciously, the stones were still in alignment – each was in relation to its neighbours, and each contained a story unique and precarious as it hung in the sea of space. Every act of violence against the seabed is vandalism against that order.

There was meaning in the stones themselves. The water had changed my mind. *You see?* I felt Rán saying. *I dragged you down there for a reason. You won't take air or stone for granted now.*

Reckon I respect the sea now, Dad, I thought. And respect meant more than fear.

My God! Seals!

Farne Islands, England

Everywhere there were soft corals, cream and yellow and orange against the blue of the water and the pink of the algae-encrusted rocks. Between the growths of coral, fuzzy white urchins were dotted. In the distance there were wrasse patrolling. The proliferation of life was astonishing, and the scene was bordered on one side by the sheer rock wall that reared from the seabed, a great wave frozen in time.

In the late summer there was great excitement at the dive club because our trip to the Farne Islands was coming up. We'd get away from home for a bit, where we had all been trapped by Covid for most of the year. More importantly, for the rest of the club, it meant the chance to dive among seals. For me, it was primarily a chance to gain experience and take my diving to another level. This trip, if it went to plan, would contain two big 'firsts': my first dive off a boat, and my first descent beyond thirty feet.

The Farne Islands are a clutch of rocky islets a couple of miles off the coast of Northumberland, the northernmost county in England, which extends far up the country's eastern side. We had all booked accommodation in the small coastal town of Seahouses, the nearest port on the mainland from which dive tours to the Farnes were operating. Its name is apt. Squat buildings pile inland from the sides of the harbour, as if they had recently shuffled out of the waves. As we entered the town a light mist blew in from the east, and the narrow lanes, stoical in the damp air, huddled under thick grey skies. The sun broke through for a while then shrugged off over the land. Evening fell and the lights in all the windows cast their restless yellow reflections on the water and, like all little quayside towns when looked upon by night, Seahouses took on the appearance of an ocean liner, beached and welded on to the coast, never to set sail again.

The sea smelt near. The scent of its life was threaded through the streets. This characteristic odour of salt water is largely caused by dimethyl sulphide, produced by marine bacteria as they digest phytoplankton. A lifeless ocean would not have this aroma. It is the smell of life and death.

By the harbourside sat a pub, the Olde Ship Inn. Its interior was crammed so full of nautical objects that actual ships would feel naked by comparison. The gel-like glass of old brass lanterns glinted on wood-clad walls. People stood around fat dark barrels drinking pints of foaming ale, holding on to their tankards as though expecting the place might list and spill their grog. Loops of plaited rope draped thickly from strong steel hooks. Harpoons lay useless but sharp on high brackets out of reach, and in the corridor, photos of long-dead fishing-boat captains stared into the eyes of customers on their way to the toilets. They looked like honest and sea-cured men who said little, went each day to work in the mouth of death, and didn't expect an easy life or a good old age or plumbing. Men who weathered the storms on deck and pissed over the side.

The guys from our club were crammed in the far corner. From phones in the palms of hands a blue-white glow shone upwards into their faces, a pale and watery light, as they contemplated online maps of the town and the diving schedule ahead. There was a sense of exotic arrival: it felt as though we were not quite still in England. Eighteen months of Covid-19 had made the country enormous, stretching the 400 miles we had travelled from home into something more akin to the great distance it would have seemed in the days before modern transport. This felt to us like a new England, and even like New England specifically as we hunkered and drank in a pub where the whalers used to sing, and where on the nearby water the boats of captains with Celtic surnames softly jounced about.

Sam, a friend from the club, said 'cheers' and clinked my pint of beer. His eyes were wide with the thrill of what we hadn't yet seen, and he grinned. He asked if I was excited, nervous, afraid. By now he'd done so many dives he couldn't confidently say exactly how many, but he had once been as green to diving as I was now: I could see in his expression the mixture of confidence in his own skills that I did not possess, and perhaps a flicker of envy that I was experiencing the steep learning curve of a recently qualified diver about to be tested in the open sea.

We got round to discussing the thing that was on all our minds: whether we would get really close to the seals, or not. There were so many factors involved: the timing of the season, the age of the pups, and the fact that they sometimes seemed more willing to approach female divers. Seals, I was told, are also happier to swim up to divers on rebreathers, who do not exhale the sudden noisy columns of bubbles that give them away more quickly as creatures who do not belong in the water. There was consensus, though, that the single biggest factor was luck.

*

The next morning came in with a drizzle that beaded every window and face in the town. We parked up alongside the harbour wall and got into our drysuits, quickly warming up in our thermal jumpers and trousers trapped under the suit's outer layer of woven trilaminate fabric or dense crushed neoprene. I considered the drysuit. My dislike for it was abating. It was a friend against the chilly summer waters of the North Sea. I also hoped my avoidance of caffeine would help me go longer between toilet breaks. It had been intimated to me that the facilities on board our boat would represent a minimalist take on the concept of a toilet. Maybe I'd manage to scrape through the next few hours without having to experience it first-hand.

We carried our cylinders, weights and gear down the slippery stone stairs that led to the mooring. When everyone was aboard – fifteen divers, the captain and the first mate – we looked up and waved at the line of our partners and well-wishers standing high above us, behind the rail on the harbour wall. Candis smiled down at me and I grinned back, waving a dayglo orange arm. Somehow I had got to this point by telling myself that I had dived a few times now and was completely unafraid. But something about that distance, between our loved ones high on the sea wall and ourselves on a small boat bobbing about by the quay, huge and dark and implacable like a castle's sides, called my bluff. *We are going to motor to some uninhabited islands, jump off a boat and sink into a cold sea.* The simpler the terms, the crazier diving sounds.

I looked around the faces of the other divers, new friends from the club and the others I'd only just met. Each meeting was polite and brief but this was my first experience of that unspoken intensity: against everybody's hope, there is the chance that something might go wrong, and then we'd be relying on each other. Each face told a separate story: there was glee, good humour, focus, the odd flash of frustration as someone wrestled with a bit of their kit. But nobody looked the slightest bit afraid. It helped me. How could this many people be about to do something if it was as absurdly stupid as diving sometimes, suddenly, seemed? We chugged away from the harbour and towards the little islets, so small that apart from their stillness they might have been the dark backs of whales. I turned around a last time. Cand and the others were tiny now: they had stopped waving and were walking off towards town. I breathed the live scent of the sea-air deeply, and put my focus aboard.

*

En route to the dive site, the first mate caught me staring over the gunwale.

'What yas looking at?' he asked.

'Uh, a bird?' I said.

Beside the boat there was a white seabird floating about in a dignified pose. It had dark eyes, a white head and grey wings, and looked like something between a gull and a large pigeon. The top part of its beak appeared to be damaged. There was a break in it halfway along. It gave it a slightly forlorn look.

The mate came over and stood beside me. On his inside left forearm was a tattoo of a beautiful woman with blue-green hair, hair the colour of the North Sea under the sun. *Rán*.

'Oh aye, you've got yourself a *fulmar* there,' he said, stressing the word 'fulmar' as if it meant 'king'. 'Lucky to see one of them, like. Aye, a fulmar. They breed on cliffs. But outside the breeding season they're pelagic birds. They keep out at sea to feed. People mistake 'em for gulls. But you can see there they've got a tube nose, a hollow nose?'

'Yeah.'

'They can smell chemicals let off by plankton. Helps navigate, like. And find their food. They'll have a bit of anything, them. Squid, fish, whatever they can find. And they can squirt this liquid out of their nose. Puts off predators. So you don't want to get too close and wind 'em up! Aye, a *fulmar* that is.'

Impressed by his knowledge, and his deep appreciation for having seen it, I stared at the bird. It bobbed about calmly, its dark eyes watchful.

Later that night I looked fulmars up. The northern fulmar's main historical breeding ground was the archipelago of St Kilda, forty miles off the Outer Hebrides, where fulmars, alongside gannets, the occasional puffin and the birds' eggs, had been the islanders' staple food. The St Kildans would abseil down the cliffs and seize the birds and their unhatched eggs, suspended from pegs that, at a distance, might be mistaken for Celtic crosses driven into the edges of the headlands: they look like little memorials to the dead, and commemorate a way of life long lost. The oil drained from the fulmars' bodies fed the flames in the islanders' lamps. Fulmar feathers lay underneath them in their mattresses as they slept. It seemed like a sad but enchanted kind of dependence: bringing warmth, fire and light to the northern darkness before the age of crude oil, before humans learnt to tap and bleed the rock. The name fulmar is of Old Norse origin. *Fúlmár* meant something like 'vile gull', probably because of the obnoxious fluid the birds would occasionally squirt from their nose-tube. In flight, the tips of the fulmar's wings look more like an eagle's than a gull's. They can live for forty years. On St Kilda the fulmars are the enduring Athenians: the humans – bar the lonely ranger – are gone, evacuated to mainland Scotland back in the 1930s, their street of houses a rain-lashed Atlantis. No wonder the first mate admired fulmars. No wonder he kept on saying their name, in which the language of those who christened the birds was briefly resurrected.

'Aye, a fulmar. When I saw ya's looking like that, I thought ya'd found the lost city of Atlantis!'

'Believe it or not, he's writing a book about that,' said Gareth, nodding towards me.

'Oh aye?'

'Trying to,' I said.

'So what's the craic, like? Is it real?'

'Depends what you mean by "real".'

He released a raucous laugh. 'Aye, you would say that! Aye! Writers, eh? Go on, give us a straight answer, man!'

'I'll come back to you in a couple of years.'

'Aye! Well, I hope you find it!'

'Me too.'

'Got to be something out there, like.'

Aye was simply the word for 'yes' in those parts, but it sounded right for another reason: it was the proper affirmative to use on a ship at sea. Every time I replied to a question with 'yes' instead of 'aye', I felt like a landsman and a feeble southerner. But the mate's last words, *got to be something out there*, were a wind in my sails.

A little while later the skipper, a broad-chested six-foot-tall man in his thirties with dark brown eyes and a smiling, bearded face, stepped out from the wheelhouse to inform us about the first dive. We'd be exploring a reef and a 'wall' – a word which, in dive briefings, means a naturally formed vertical rock face, and not the sort of walls I'd started diving in hopes of seeing. Still, it sounded enticing. We could expect to see plenty of coral growth, crustaceans and fish, and we might get lucky and catch sight of some seals.

'Remember to put your SMBs up when you're ready to ascend, all right?' said the skipper. 'Then we'll have a chance to get over to you by the time you're up.'

SMB is short for 'surface marker buoy', a brightly coloured collapsible buoy which can be filled with air and sent to the surface to let boats and other observers know the position of a diver below. Some divers call them 'bags' for short. You have to take great care when sending a 'bag' up from depth. If it gets caught on your equipment or the line becomes tangled or jammed, it might take you rushing up with it.

'Once we see your bag, we'll get to you,' continued the skipper. 'So, this dive: you'll descend immediately to about fifty, sixty feet and follow the rock face round the corner. Simple. Most likely you'll finish on forty-five or fifty minutes.' I gulped. Sixty feet was twice as deep as I'd ever dived before. It might not sound like much but there's a reason the basic open-water limit is set at that depth: it is the deepest you can go while 'no-stop' diving, which means you can surface at any time without having to plan for safety or decompression stops on the way back up. There's plenty that can go wrong above sixty feet, but it's the point at which diving suddenly becomes more complex, a sort of second surface beyond which scarier rules begin to apply. The next invisible horizons are 100 feet — the limit for more advanced recreational divers — and 130 feet, which is the cut-off depth for recreational scuba: the frontier of technical and professional diving. As I checked over my regulators and gauges before the jump, those depths seemed impossibly far off to me. Sixty feet was daunting enough. We were about to drop the height of a seven-storey building.

I was going to dive as a trio with Kenny, with whom I'd now dived a couple of times in the shallows off our home beach, and John. I hadn't met John before. He had maybe fifteen years on me, and with his silver goatee, focused stare and close-fitting drysuit he looked like a seasoned diver with a slightly military aura. He met my eye across the deck and gave me a serious nod. I nodded back. I was putting a brave face on it.

We got into position near the stern of the boat. The sea had a skin of dark jade. We stepped off in turn and hit the water. After signalling thumbs-down we descended. John dropped fast. Kenny went after him. I stalled. I'd let all the air out of my BCD but I wasn't going down. I emptied most of the air in my drysuit out through the neck seal: it left past my chin with a warm huff. My legs were tight in the suit now: it gripped unforgivingly under the water pressure. I waited to descend. Nothing. Was I underweighted? Possibly. Or perhaps my lungs were fuller than usual because I was nervous. I looked down beneath me. John and Kenny had almost reached the bottom. I could see them fairly clearly through a pale blue mist, beside their rising fountains of bubbles which were shimmying up towards me. For British waters the clarity was astonishing.

I decided to try and get down with concentration, and a big push. Diving forwards duck-fashion with a curl, I straightened out and kicked my legs. It worked. I began to drop and was soon at fifty feet. A quick look up confirmed how different boat diving was from the shore dives which were all I'd done until now. There was no reassuring sense of a beach being nearby. Beside me a steep rock face led up to the waves, and below, the reef stretched off like a valley on an alien planet. There was no path back ashore along the bottom. The only way out of here was back up, and on to the boat. I added a little air to my suit to offset the squeeze. Perfect. I was comfortable now, and neutrally buoyant, weightless in the cool North Sea.

My next thought went to my buddies: confirm we're all fine and heading in the same direction. Kenny was checking his gauges. He saw me and gave me an OK. John was up ahead. He turned around and his eyes were wide: for a second I wondered why. Then I took in our surroundings. We were on a true cold-water reef. A vista of pastel colours stretched ahead to the limits of our vision: the yellow and cream of soft corals, the deep pink of algae on stone. Beneath a ledge I saw a blue lobster sheltering from the light as if from rain, his long antennae dowsing for information. Kenny and John were sending plumes of bubbles up from their regulators, pewter-coloured in the misty blue. It was the first time I had been deep enough to appreciate the beauty of these inverted triangles of rising air. I watched them against the rainbow backdrop of the reef, the pigments of chilly seas. As a diver moves forward, releasing a breath every five or ten seconds, glinting trees of bubbles rise up behind them in a climbing sequence before disappearing forever. They looked like a painter's allegory: the passage through life as a series of breaths. I took out my camera; even with my limited skills this footage was bound to look spectacular. *Wait until I show everyone back home the kinds of places I have visited*, I thought: off the shores of our own country, and yet more bizarre than any place on land even if it were 10,000 miles from here. I checked my computer again. Sixty feet. I was operating at my limit, and thrilled to be there. I grinned and took a selfie. I was doing it. Diving was no longer a pipe dream.

The excitement must have been affecting me, as I started to take deeper, faster breaths. I rose slightly in the water. Then I saw that John and Kenny had moved further ahead. I needed to make up some distance, and breathed deeper still as I put some force into my kicks. But I wasn't just moving forward. I was moving diagonally upwards. And I was gaining speed. We must have advanced over a shallower stretch of bottom and, overwhelmed by the stunning scenery, I had failed to keep a close enough eye on my buoyancy. I dumped some air out of my BCD but by this point I had reached thirty-three feet. *Oh no.* My buddies were far away now. I focused my mind downward, attempting to control the ascent, but as my breathing sped up in response to the situation I rose even faster, and the little air I had let into my suit at depth had now expanded and I was far more buoyant. Twenty-five feet. I was now closer to the surface than the bottom. By the time I forced out all the air from my drysuit I was barely ten feet below. I had climbed fifty feet in less than a minute, barely within the parameters for the maximum safe-ascent rate. The sea's warped skin twinkled above me. I had lost sight of Kenny and John. If I went back down I'd be unlikely to find them, and besides, we had been separated for more than a minute and a half. I only had one option now: to wait up top until John and Kenny emerged or the boat came over to pick me up.

So this was why they restricted new divers to a depth limit of sixty feet. Teresa had warned me about the risks of taking a camera below when I was still inexperienced. There was a danger of task-loading: losing track of your priorities, concentrating on getting pictures instead of keeping your mind on your buoyancy. I had thought I'd be fine. It turned out Teresa was right.

Defeated, I broke the surface and filled my jacket with air as small dark waves slopped around. My computer beeped a discordant tune and flashed yellow warnings, notifying me that I had failed to do a safety stop. I knew I was OK because I'd been down so briefly, but it was still demoralising to get an audible warning from my equipment. In the space of three minutes I had gone from an elated man, drifting in a state of total aesthetic contentment over an enchanted submarine world, to an isolated novice who had popped up way too soon from his first dive off a boat. I rode the swell and stared at the little islands ahead. They looked slate grey and forbidding. From the rocks at their near edge a few seals stared at me. I felt lonely and inept. Kenny and John would not be happy I'd caused them to bail early on the first dive of what for them was a much-anticipated holiday. They'd spent good money on it. They'd been kind enough to buddy up with the new guy. And I'd repaid them by disappearing on sixteen minutes in a barely controlled ascent. After an escapade like this, surely they wouldn't want to dive with me again.

The boat came over before Kenny and John resurfaced.

'Everything all right?' asked the skipper.

'Yeah, fine,' I lied.

'You've not put your SMB up.'

'Oh. Sorry.'

'I love it when people listen to my dive briefings!' He shook his head.

'Apologies.'

'Why'd you come up?'

'Sorry. Weight problem.' I didn't want to say I'd experienced something uncomfortably close to a runaway ascent.

'Aye. Well, let's get you back aboard.' He could see I was unhappy. There was no suggestion I might wait for my buddies and try and go back down.

John and Kenny appeared. Considering how quickly I'd left them behind they had found me in very good time.

'What happened to you?' asked Kenny. 'We looked around and you were gone.'

'Sorry, guys. I think I was underweighted. I had problems staying down.'

'Oh, aye,' said Kenny. John said nothing. I felt like I'd blown it with him on our very first dive.

'I couldnae get down at first myself,' Kenny said. 'I'm going to add some more weight on the next one.'

It was a window of forgiveness. I felt a bit less stupid now. And this was new territory for me: my first time outside the shallows in a drysuit. I tried not to beat myself up. But we should have been down there for longer. I'd deprived myself, and more importantly my buddies, of some precious experiences; of memories. Maybe it was forgivable, but I was determined not to let it happen twice.

We dived once more that afternoon: Kenny stayed with me but John, understandably, opted to go down with another pair. I left my camera behind and focused on my buoyancy, to the benefit of procedure but to the detriment of my enjoyment. Blue velvet swimming crabs stared at us from beneath dramatic ledges of rock, and huge wrasse swam close by our heads in hope of a meal: other divers, I was told, would sometimes smash sea urchins to entice the wrasse closer for photographs. I struggled to appreciate them, too worried about losing control again. We surfaced after a long and well-executed dive, but I was still upset about spoiling the morning one.

*

Back ashore, I rang around various numbers in and outside town, trying to find someone who would sell me some lead weights at short notice. I eventually got hold of a diver who sold equipment from a lock-up on an industrial estate on the edge of Seahouses. He turned up in a truck outside his unit and threw up the grille.

'Who you diving with?' he asked.

'Billy Shiel,' I replied.

'What's the rush for the lead?'

'Ah, we did a couple of dives today and I was a little underweighted.'

'Aye, you don't want that. Not when you've come all this way, like.'

'No.'

'How much you wantin'?'

'Hmm. Twelve or fourteen pounds?'

'You can't have been that underweighted, man!'

I explained that my buddy might want a couple of extras as well, so it was just to be on the safe side. I went for two 4½ lb blocks and a couple of 2 lb bags of shot.

'Bloody expensive, lead, mind,' he said. It was true. A day's wages for four small weights. No wonder people steal it.

'I'm just happy you were about,' I said.

'Trouble with diving,' he replied. 'All the gear. Pain in the arse, like!'

★

The history of diving is almost inseparable from the development of its equipment. This is not only true of modern practices with scuba and rebreathers, but also of ancient forms of diving. The first man or woman who held on to a stone to help themselves sink to the bottom with less effort, thereby making their breath last longer underwater, had moved off the path of the seabirds and otters, who went below the surface equipped with nothing but their own bodies. From the simplest physical diving aids to the dive computer, which keeps track of a dive's depth and duration while estimating soft-tissue nitrogen absorption, these are all enhancements of the human into a being better suited to staying under the water for longer. They are replacements for bodily evolution, a shortcut to powers that nature would take millions of years to confer; or, in another perspective, restoring abilities that nature has, over millions of years, swept away. The human body, too, is a ruin, its old form smoothed and altered by the weathering of genetic change.

An exposure suit, essential as they often are, is only the skin of the matter. The real miracle is to draw a breath. It's a transgression of our animal limits. To ferry air down below the surface, to smuggle a pocket of sky down with you, is a Promethean undertaking. It turns air into elemental contraband. As a raw feat of endurance, breath-hold diving will always be more impressive than scuba diving, but there is something mischievous about actually inhaling breaths while submerged, and while free from a cable to the surface: a tether to the terrestrial, a leash to the human domain. The scuba-diving pioneer and writer Philippe Tailliez caricatured human history as a voyage of escape from 'cables', and the invention of self-contained diving equipment as one of the great leaps on this journey. It finally enabled us to move underwater with the freedom we had always had on land and which, during the first half of the twentieth century, we had been discovering in the sky. For Tailliez, the genius of scuba equipment was not that it enabled humans to breathe underwater – surface-supplied divers had been doing that since the early nineteenth century – but that it made the source of air portable. The diver had been set free from the tether of a breathing hose. At last, they could swim about unleashed. The breath now followed the will.

When the scuba diver, unshackled from the rest of the human world, swims while breathing underwater, he or she is doing something that is almost impossible to find elsewhere in the animal kingdom. Creatures that breathe underwater are the rarest of the rare. The Costa Rican lizard *Anolis aquaticus* – the water anole – is one example. In recent years, the naturalist Lindsey Swierk has been amassing evidence that the lizard can retain pockets of air on its skin when it dives, and even attach a large bubble of air to its head, breathing from it while submerged as though drawing breath from a skinless balloon. In this way, the *Anolis* is able to dive for up to sixteen minutes, leading it to be nicknamed the 'scuba-diving lizard'. Photographs of it always look slightly fantastical: because of the glassy, silvered appearance that a bubble of air has underwater, it looks like the lizard has a glass bell jar attached to the top of its head, a sort of transparent policeman's helmet, giving it the bearing of an eccentric – and reptilian – nineteenth-century inventor.

In the urtext of scuba diving, Jacques Cousteau's *The Silent World*, we are introduced to the marvel of a different sort of aqualung – the mechanical, man-made kind – as it felt to the author and co-inventor. The book was published in 1953, while scuba diving was still a new, exotic, dangerous and, for most people, inaccessible activity. Cousteau and his companions were so close to the cutting edge of these new contraptions that on several occasions they almost lost their lives testing them. They dived with early rebreathers, closed systems with no air output that used caustic solutions to 'scrub' the carbon dioxide from the air they breathed out, allowing it to be 're-breathed'. The concept of a rebreather that removed CO_2 from exhaled air had a long history, dating back to Napoleon's Imperial Navy in the early 1800s, but refinements had to be made before they saw widespread uptake. By the 1870s, the English diver and engineer Henry Fleuss had invented a rebreather system which was practical enough to see some commercial use, a 'self-contained breathing apparatus'. Its rubber mask, which had two oval eye-holes and a pair of breathing tubes that ran down the chest to a set of pipes, looked like the face of a mutant elephant. Fleuss's system worked by supplying a high-oxygen gas mixture from a copper tank, and 'scrubbing' the excess carbon dioxide from exhaled air by passing it over a rope soaked in a caustic alkaline solution of potash, or potassium hydroxide. Like all early rebreathers, it was a basic system without the capacity to fine-tune the release of oxygen, meaning it was only suitable for depths above 130 feet where oxygen toxicity presents less of a risk. In spite of this limitation, the rebreather became the standard choice for military divers – or 'frogmen' – in the years up to and including the Second World War. Open-circuit diving equipment, for all its benefits, would always present the fundamental problem of releasing bubbles, which for an underwater soldier are the equivalent of lighting a signal flare to declare your position to the enemy. Reading about how early rebreathers worked, though, I wondered whether I'd have been more terrified of the rebreather itself than of hostile eyes or guns.

In any case, underwater, making a new discovery can often mean a brush with death, as when Cousteau was experimenting with breathing pure oxygen before the dangers of its toxicity at depth were known: more than once he blacked out having just about managed to ditch his weights, before floating upwards and being coaxed back into consciousness at the surface. But perseverance paid off in the end. The aqualung promised to deliver on the dream of being able to breathe below the surface, while largely unencumbered and untethered by lines to a boat. At the start of the book, Cousteau tells how

> One morning in June 1943, I went to the railway station at Bandol on the French Riviera to collect a wooden case expressed from Paris. In it was a new and promising device, the result of years of struggle and dreams: an automatic compressed-air diving lung conceived by Émile Gagnan and myself.

Here, provided by science, was the techno-selkie-skin. Engineers had delivered what ancient legends had merely promised. A self-contained underwater breathing apparatus – a thing now classed as a holiday experience and spoken of quite flippantly in terms of hobbies and pastimes, of sport – was for Cousteau a vehicle out of his imagination and into the heart of the sea: a world where biology and physics were melded with dreams.

The writings of early recreational scuba divers often took flight into a kind of speculation that might seem pretentious, or even mad, to most modern divers. In the late 1940s, Robert Gruss, author of *The Art of the Aqualung: How to Swim and Explore Underwater*, claimed that:[1]

> Diving in self-contained equipment has created a new race: Men of the Sea. They view it in its totality, they bear its weight and try to learn its secrets. I have never considered diving as an ordinary distraction, or even as a sport. The moment the sea closes over me I feel some great thing is happening. I am filled with a kind of awe, without really knowing why.

Philippe Diolé, a contemporary of Gruss, wondered whether

> the ocean depths give us the chance of a new humanism. Yet I think we have not quite reached that stage; I'm not yet very happy about the word. Is it really a humanism, this slow impregnation, this gentle and pervasive enlightenment, prerogative of the 'men of the sea'?

The complex-looking equipment that scuba diving requires has enabled divers to enter a realm where apparent simplicity, the simplicity of three-dimensional space, invites new ways of thinking. What exactly is the 'slow impregnation' Diolé refers to? The word 'dive', traced back through its Germanic roots, leads to a Proto-Indo-European origin-word that probably sounded something like *dheub*. It would also have been the root of the word 'deep', and had the sense not just of depth, but of hollowness: of cavernous depths, or the spacious depths of the heavens. Diving into water or air is possible because they lack solidity. Perhaps myth or legend are similar in that they lack the fixed connection to measurable events, to chronology and archaeology, that can obstruct the flight of imagination.

The enchantment of Atlantis is akin to the draw of the sea, or the beckoning of the wide air that the parachutist feels: it is not just what can be seen that beckons, it is the emptiness as well. Chock-full as Plato's tale of Atlantis is with detail and writerly ingenuity, it is also crucially lacking something: a tether to a particular locale, which would have prevented so many minds, speculative and practical, brilliant and mad, from leaping into it full of wild abandon.

As I drove back to the hotel to prepare my things for the boat, I thought about the guy who had sold me the lead. He was a diver, a long time after the 'golden age' when it was regarded by its practitioners as a miracle, a mystical experience, an absurd privilege wrought by engineering: the underwater sublime. *All the gear. Pain in the arse, like.* What would Cousteau, Tailliez and Diolé have made of that?

★

The next morning we were told there was a chance we'd see some seals, but also reminded that 'Ultimately they are wild animals, so nothing is guaranteed.' The pups had been in and out of the water which often augured well for the chance of 'close seal interaction'. Kenny and I dropped side by side but this time the entire group stayed closer together. A loose cluster of us hung in the water, finning slowly past a high wall that reared up off the seafloor. I'd added some weight and felt good in the sea: heavier, but relaxed in the certainty that the extra few pounds would prevent me rising too fast. I felt able to use my camera again. If I had any issues with the dive I resolved to pack it away immediately and focus on procedure.

We swam forwards past an underwater 'clifftop', the edge of which was frilly with swaying tan-coloured fronds of kelp. Something dropped from one of the fronds and fell slowly, like a deflated balloon. Its colour matched that of the kelp. I noticed it had its own fronds, curled neatly behind it. They unfurled. Tentacles: it was a common octopus. I was so pleased to see it I forgot my nose was blocked by the mask and my head jerked as I tried to take in an impossible breath. The octopus stretched out its arms behind it like streamers and arrowed away into the cloudy blue beneath us. It moved with swooping elegance, as if releasing its position, rather than darting towards another one. It is always a privilege to see an octopus. Nothing else in the water moves like that, with such a smooth blend of caprice, panache and speed.

I looked to my side for Kenny. He gave me an OK. It felt like yesterday's mistake had been forgiven. I 'OKed' him back, then looked to my side at the huge wall of rock. A shadow moved. Then there was a silver flash.

Down a vertical crevice in the stone there swam a large grey seal. Its fur was lustrous in the water. Its movement was slicker than anything with bones should be able to manage, and it moved to the bottom of the cleft of rock like mercury easing through water. At the bottom it turned and swam back up, its passage suggestive of curiosity or exercise as opposed to a search for food or a flight from danger. It was maybe twenty feet away, closer than I had expected a seal to come to an ungraceful diver like me. When it passed over the wall and out of sight I looked at Kenny and shook my head in a 'How can something that wonderful happen?' type way. Kenny gave me another OK, and this time the signal seemed to mean 'Everything is all right with the world'. Almost as astonishing as the beauty of the seal and the octopus was the nuance with which divers managed to imbue the simplest gesture of a hand.

★

The next dive took place mostly in the shallows as we stalked through the kelp, hoping to catch sight of some seals. I browsed the weeds, brushing them aside with my fingers to see what might be hiding there. I disturbed a spider crab. It had been perfectly camouflaged by straggly fronds of seaweed attached to its back. As it scampered away its shell resembled a Second World War-era Brodie helmet disguised with a net and green tassels, except in this case the disguise was itself alive. The crab was hidden not by masquerading as something else, but by letting itself be a home to other things; by embracing other life, it was protected. Likewise, the school protects members of a single species by diminishing the chance that any one individual will be seized. Schooling – the coordination of motion, trajectory and speed – provides each fish with the eyes of its fellows and might also help with hydrodynamic efficiency. The crab smothered in other forms of life represents a kind of schooling across the borders of species. The ribbons of seaweed, the grasping hydroids or anemones stuck to its shell, wafting in different directions, give an overall impression that this cannot be a single life form on the move. It looks like an environment, a little world.

This area was known as Gun Rock: apparently there were cannons from an old shipwreck nestled among the kelp in the flat areas. Theoretically I should have been more interested in finding these human relics than the seals, but my heart had other ideas. To our right were large horizontal ledges of stone. A slight current helped us drift past them. I shone my torch under one and the electric-blue trickles of pigment on a squat lobster lit up in the beam. I rose up parallel with a ledge. Then I took a double take: lying atop the ledge sideways, eyes closed and seemingly asleep, was an enormous seal. A male, judging by its size. The current was rocking it slightly and the kelp fronds dangling down near it looked like the ragged curtains of an undersea bedchamber. I was astonished and a little afraid. It was less than six feet away and much larger than me, a chubby bulb of life dozing in the water. And it was a mammal. In an instant the old ideas of mermaids and selkies stopped seeming ridiculous. There was something in the seal that was still of the land: cow-like, dog-like, human-like; something breathing and warm, whiskered, wearing skin, not scales. Then I understood there was another reason why I was moved to be so close to it: as the seal slumbered it held its front flippers crossed on its chest. My dad used to sleep like that. Of all the things I guessed I might experience diving, I never thought I'd meet something with body language that reminded me, in such a precise way, of my father. That an animal of the sea could be, even if in a limited way, a dead ringer for a human. For a parent. For family.

It opened its eyes. The huge dark marbles blinked a few times then looked at me with a benign curiosity. As it angled its head to get a better look, folds appeared in the flesh below its muzzle: 'chins' of fat, as a human or puppy might have. How absurd it was to not regard this creature as a person, kindred. Its personhood was encased in a shape adapted to its environment, but why should that lead us to conclude it wasn't a person? Wide awake now, the seal wiggled off the rock and disappeared around the corner.

Back aboard the boat, I offered bars of chocolate around. I had a few takers, hungry for calories after an hour in the cold sea. I slumped against the gunwale and let my hair down: it was all over the place, salty and dank as seaweed.

'Someone's had a good dive!' said Angela.

I nodded. Angela smiled. One more student of hers had come to understand.

Another diver, an underwater photographer wearing a rebreather, came aboard. He looked like he'd won the lottery. He had.

'Nine! There were bloody *nine* of them! All around us. Un-bloody-believable. My God! Seals!'

'Happy, then, are we?' asked the first mate.

'Happy? That was crazy. They were so close. Not one. Bloody nine!'

The deck was awash with contentment. Until recently I hadn't even known people dived to see wildlife in UK waters. It was something they did in Bali, Thailand or on the Great Barrier Reef: the famous scuba and snorkelling sites where life thrummed in abundance. But in this moment you couldn't have convinced any of us to swap places with divers anywhere else in the world. We had just swum with seals. We had shared their water as we witnessed the liquid perfection of their style, ribboning around us, tassels of life, the silver dogs of the ocean.

And I was happy for another reason. In fixing the weight issue, getting down easier, it felt like I had added some gravitas to myself. *You had a problem at sea and you solved it by adding more weight. You didn't solve it by getting out. You solved it by going down heavier.*

I didn't know whose voice this was – Rán's, Dad's, an amalgamated voice of my diving tutors, or my own. Perhaps all of these voices were blending, mixing into what, for me, was the voice of the water.

Transfer to Reality

Hasankeyf, Turkey; Lion City, China

With a long breath
and unwavering conviction
you became the stone cradle of humanity.
In the light of the sun
and the protection of the darkness
you bore children tirelessly –
Amed, Heskiv, Ninawa,
and Gilgamesh
nurtured them
into flowers of civilization.

Now they want to steal the tongue of Amed
and drown Heskiv in your belly
but your old power
and your young fury
feed the resistance of your children.
How long is thy breath,
Tigris river, the cradle of humanity?

Zaradachet Hajo[1]

Birthplace of Gilgamesh or not, the city of Hasankeyf was drowned by the Turkish government's Ilisu Dam project in 2019. Continuously inhabited by humans for over 2,000 years, and perhaps as many as 3,800, it had survived the rise and fall of no fewer than nine distinct civilisations, including the Medes, Romans, Byzantines, medieval Arabs, Mongols and Ottomans; but it could not survive the rising waters brought by hydraulic engineering. By April 2020 the entire old city was underwater. Years of protests had failed to save it. Descendants of its residents, should they wish to see where their ancestors lived, will have to dive down to them. Many believe that securing the regional water supply was not the government's only motive for flooding swathes of this part of Turkey. Hasankeyf lay in the south-western corner of Turkish Kurdistan, and its submersion worked as both a physical and symbolic drowning of a Kurdish cultural stronghold.

Efforts were made to move some important structures uphill to where the rising waters would, supposedly, not reach. The care taken over the exhumations – of long-dead nobles and recently departed family members of the locals – seems somehow perverse. Kid gloves were donned in order to delicately re-situate the dead, following a government decision to flood the homes of those still living. Even religion was forced to take the road uphill in order to survive; a giant spatula slid under the mosque to lift it in one piece on to its transport.

I checked online maps of Hasankeyf before and after the Ilisu Dam. The digital version of the River Tigris was constantly updated as the reservoir grew deeper and wider by the week. It was a record of a drowning in progress; a fall of Atlantis largely ignored by myth-chasers because it was happening not thousands of years ago, but in their own time, shrouded in disregard. It made a peculiar counterpart to the time-lapse images of the Aral Sea, the once-giant lake between Kazakhstan and Uzbekistan that has shrivelled away into a desert strewn with shipwrecks. The reservoir moves in the opposite direction, though it is driven by the same forces: technology and human will.

'The city and citizens which you yesterday described to us in fiction,' says Critias in Plato's *Timaeus*, 'we will now transfer to the world of reality.' With the Ilisu Dam project the Turkish government had taken the tales and made them true. Hasankeyf was their real Atlantis.

*

Were Plato's Atlantis story literally true, then the knowledge of the fall of Atlantis would have come to us, the readers, via at least ten 'stages' of transmission. It must have been conveyed by a hypothetical eyewitness to the event, to a recipient in Egypt, and thence through 9,000 years of tradition to the priest at Sais who supposedly related the tale to Solon the Lawgiver, who brought it back to Athens. Solon then told the story to his friend and relative Dropides, who told it to his son Critias the elder, who told it to his grandson Critias the younger, who mentioned it at a meeting with Socrates and others. Someone at that meeting then must have told it to Plato, who finally wrote it down, thereby telling it to us. The reader of the *Timaeus* and *Critias* is the tenth step in the transmission of this information. That being said, one of these 'steps' is a 9,000-year priestly tradition (itself, therefore, composed of an unknown number of 'steps') which not even the ancient Egyptians ever claimed existed. In short, the provenance of the Atlantis story is perhaps the longest and most improbable tale of pass the parcel in the history of human storytelling.

One of Plato's greatest achievements with the Atlantis story was his ability to make some of his readers believe in it, in spite of the ludicrous way it supposedly came down to him. It may even be that the complexity of its purported origin somehow makes it more believable. Counterintuitive as it might seem, it could be more tempting to believe an intricate cock-and-bull story than a simple one. Plato's use of Egypt in the story might also have helped to suspend his audience's disbelief. The antiquity of Egyptian culture, its mind-boggling age and majesty, casts the dust of truth over the story, bringing Atlantis onstage already dressed in the garb of a place that definitely existed. Plato also starts his yarn in the domain of the real, with Socrates and his student Critias,[2] real people whose lifetimes overlapped with those of Plato's first readers. From Critias he moves to Solon, who had been dead for about 200 years by the time Plato was writing. But Solon was also an undoubtedly historical figure, and based on the writings of Herodotus, he really had spent time in Egypt.[3] The fictional story of the war between Athens and Atlantis is told 'in parentheses' within a wholly believable tale of a real man visiting an actual country. It's like placing a forged painting in a gold frame or putting artificial flowers in an expensive vase.

This cunning narrative trick, which seems to have been invented by Plato, has been performed plenty of times in the twenty-three centuries since. Samuel Taylor Coleridge used it in his long poem *The Rime of the Ancient Mariner* (1798), in which a wedding guest is accosted by an old sailor: his tale of the albatross and the cursed ship occurs within a more plausible account. The technique is repeated in Mary Shelley's *Frankenstein* (1818), Emily Brontë's *Wuthering Heights* (1847) and M. R. James's *Ghost Stories of an Antiquary* (1904), among many others. In *The Lord of the Rings* (1954), J. R. R. Tolkien tells us that the story came down to him in a manuscript called the Red Book of Westmarch (an example of the 'found manuscript conceit'), a conscious allusion to the Red Book of Hergest, a very real fourteenth-century Welsh manuscript. Tolkien had read the *Timaeus* and *Critias*, and states in his published letters that his fictional island of Númenor is directly based on Plato's Atlantis, so it should not surprise us that he might have drawn inspiration from Plato's techniques as well as his story – from his palette as well as his paintings.

★

Nations too draw inspiration from each other. In response to criticism of the drowning of Hasankeyf, President Erdoğan of Turkey protested that many countries have drowned old buildings in the pursuit of modern goals.

In China's Qiandao Lake – 'the Lake of a Thousand Islands', picturesque blobs of green and gold atop a turquoise mere – are the remains of the ancient city of Shicheng, 'Lion City'. On those days of the year when the water is warm enough to permit diving, a few people, fin-footed and darkly clad, will swim down to its ruins, where wild-eyed grimacing lions of stone stare out through dusty beams of light. The swimmers make a beeline for this 2,000-year-old complex of temples and palatial rooms, dotted with masterfully carved sculptures of fierce, entrancing beasts. But all around – for miles – are other, more recent dwellings, all drowned at the same time as Shicheng's ancient halls.

Qiandao Lake is man-made. In 1959 the Chinese government finished building the Xin'an River hydroelectric station, flooding the valley behind it along with its hundreds of farms and villages. Deep-rooted communities were evacuated to make way for water, power, fish, the future. The lake's many photogenic islets were, within living memory, the tops of hills where children shouted and ran. One of them is named 'The Island to Remind You of Your Childhood'.

In Dubai, another place where engineers constantly scheme to cajole and redirect the water, Atlantis has acquired a simpler meaning. On the seaward edge of the city is the Palm Jumeirah, a miniature man-made archipelago that protrudes out like a gigantic fossilised ribcage into the Persian Gulf. On its outer edge sits the enormous luxury-hotel complex of Atlantis, The Palm. With a private beach and promises of 'other-worldly luxury', the resort channels thoughts of Atlantean grandeur and ignores any warnings of hubris. Calling a waterfront high-rise hotel by the name of 'Atlantis' tempts fate.

The Floating Rocks

Santorini, Greece

Somewhere above us the sun still blazed but at these depths it was invisible. There was light, but a light that appeared to have no source, as though the seawater itself were luminous, a blue glow emanating from every part of it. Small groups of saddled sea bream swam past, their little upturned mouths making them appear glum but resigned to keep swimming. They seemed to be lit from all sides at once, the diffuse radiance flashing in sequences off their silver scales. The world was aglow.

In the summer of 1619 BC, it was unusually cold in China. In the lowlands, snow fell in 'the sixth month', and there was frost on what should have been mornings of pleasant warmth. Crops grew more slowly than usual. The harvests were scanty and late. Those looking up at the sky observed that it now had a yellow hue, and a mist hung in the high air, in the unreachable heights between the mountaintops and the sun. Where the mist and the cold had come from, none could guess, other than that they might have blown in from some unlucky place.

These ill omens were noted by scribes. The original records are long lost, but we know of them because they were copied in later centuries on to *jiandu* – narrow strips of bamboo, each the width of a pair of chopsticks – bound into mats held together by hemp string or thin leather cord. The mats were rolled up and stowed away, some of them in the palatial tombs of kings. There they sat silently through the years, unread but enduring.

The yellow skies were likely caused by a volcanic explosion on the island of Thera, nowadays better known as Santorini, a place so far to the west that its implication in the Chinese weather seems ridiculous. The Santorini volcano is the most active in all of Greece and, according to Georges Vougioukalakis, an expert on its history, 'one of the most violent caldera-volcanoes in the world'.[1] In around 1600 BC it erupted.[2] All traces of life on the island – human, plant, animal – were obliterated. Not a seedling or spider survived. The volcano's reach proved lethal even hundreds of miles away. In 2021 the skeleton of a young man was found at a dig site in Çeşme Bay on the western Turkish coast, his body smashed against a retaining wall, a posture in which modern tsunami victims are often found. He had been killed by a seismic wave caused by volcanic activity. At a distance of almost 150 miles, the eruption had taken his life.

The Santorini volcano has flared at several points in the last quarter of a million years, each explosion re-forming the island, reshaping its cauldron into a brand-new monument to the power beneath the earth. The 'Minoan eruption' was probably the most devastating volcanic event of the last 10,000 years. It was violence on a godlike scale, the kind that obliterates a sophisticated island civilisation overnight. The kind that destroyed Atlantis.

Santorini lies on the South Aegean Volcanic Arc, a curved line of ruptures in the Earth's crust that stretches more than 250 miles from the peninsula of Methana in the Peloponnese, as far as the island of Kos in the Dodecanese, just off the western coast of modern Turkey. It is an ancient crescent of destructive potential, a sickle of Hephaestus, the Greek blacksmith god of volcanoes.

The theory that Santorini might be the 'real Atlantis' is based on a small number of details in Plato's story that map on to the island and its history. Atlantis had a sophisticated building culture long before many other parts of the world; there were multistorey dwellings on Santorini more than a thousand years before the golden age of

Athens. Plato mentions that the architecture of Atlantis was built from stones of three colours: black, white and red; this *tricolore* is to be seen everywhere in Santorini. Atlantis had hot and cold springs; so does Santorini. Atlantis's civilisation was destroyed in a single night and a day; so was Santorini's. This has led to a belief, manifested in a museum on the island called the Lost Atlantis Experience, that Santorini might have inspired Plato's creation in a fundamental way; that folk memories of it trickled through 1,300 years before making their way into the mind of Plato and out again through his writing.

The historian Tom Holland has described the Santorini–Atlantis theory as 'posh pseudo-archaeology':[3] it is yet another, ultimately silly notion about Atlantis being a real place, dressed up as something more intellectually credible. Pierre Vidal-Naquet would have agreed with him. As both of them have argued, there are far more differences between the Atlantis of Plato's story and Santorini's Bronze Age civilisation than there are similarities. Atlantis lay far from Greece, in the Atlantic; Santorini is in the southern Aegean, in the heart of the ancient Greek world. Atlantis had a great and powerful navy, and a mighty empire: ancient Thera, prior to the Minoan eruption, had neither. Atlantis was destroyed by an earthquake and it sank; life on Thera was obliterated by volcanism, and much of the island stayed above sea level. The list goes on. To call Santorini the real Atlantis is as preposterous as giving that title to any other place in the physical world. In spite of this, the idea that the volcanic destruction of the island does in some sense lie behind Plato's story is one of the more respected, or at least less mocked, hypotheses about its roots. This is mostly due to the fervour of one man, the Greek archaeologist Spyridon Marinatos.

In later life Marinatos had hair that was silver at the sides, and his heavy brows lent an intensity to his stare. On digs he would sometimes wear a pith helmet and a khaki shirt with buttoned epaulettes, giving him the appearance of a rogue colonial officer. He had been born in 1901 on the island of Kefalonia in the Ionian Sea, west of the mouth of the Gulf of Corinth. In the year of his birth, riots broke out in Athens over the issue of whether the official language of Greece should be the Demotic dialect of the people or a restored form of ancient Greek called Katharevousa or 'purified Greek'. The country was envisioning what it would be in the light of what it once was, and the visions clashed and sparked.

In his twenties Marinatos worked at the Heraklion Museum in Crete. There he met Sir Arthur Evans, who had led the excavations of the Minoan palace of Knossos, which in its time had been painted by a layer of ash that came south on the wind from the Santorini eruption.

By his thirties Marinatos was director of the Antiquities Service of Greece. He helped uncover the battlefields of Thermopylae and Marathon, the two most famous land battles in the nation's history. In the 1960s he led the discovery of Akrotiri, the 'Greek Pompeii', a highly sophisticated Bronze Age town on the southern coast of Santorini, eerily preserved by the volcanic ash that had suffocated it. He died on the site in 1974, killed, according to one account, when one of the ancient walls collapsed and crushed him. Marinatos was, arguably, the last man in Akrotiri to be killed by the disaster of 1600 BC. Death incorporated him into his obsession. Apparently he was buried somewhere near the site, but I couldn't find an exact location. I'd just have to keep an eye out.

As our plane came in to land at Santorini Airport, the islands — there are four of them in this tiny cluster — were twinkling away in the darkness. Like other archipelagos, Santorini gives the impression of being a world unto itself. Cand and I arrived at our hotel, which demurely called itself 'Paradise Resort'. It was unfeasibly clean, as though we were the first ever guests. There was a chlorinated pool with a fat palm tree growing in a notch in its side. On the tiles that skirted its edge there were occasional flashes of grey: the shadows of small lizards on fly patrol.

The volcanic ash and pumice deposited on the island by the Minoan eruption lies all over Santorini's surface, in a parched white layer which in places is forty feet thick. It is flecked with chunks of black rock that would have landed flaming among it, like mortars dropping from the sky. Beneath lies a brown layer, the 'palaeo-soil' that was farmed by the island's inhabitants before the volcanic eruption. And on top of all this sits the modern soil of Santorini. It is thin and scabby, the Assyrtiko vines and the wild and hardy desert plants of the island clawing into it for dear life. Santorini's geology is a message of stone that floats on its history. Life was here before; it ended; life is come again.

We hired a quad bike, a favoured way of getting around. At the bottom of a winding southward road we arrived in the excavated city of Akrotiri, where Spyridon Marinatos had met his end. We were greeted by a harsh smell of cinders. The place still smelt of the eruption. It was shocking that it had left such a potent tang in the air, a scent 3,600 years old but still hot and strong, like ash just swept from the fireplace. Perhaps I shouldn't have been surprised. Volcanoes keep their own time.

The ruins of ancient Akrotiri are protected from the weather by an enormous purpose-built roof, and an elevated walkway enables visitors to stroll among the uncovered buildings without the risk of damaging them. The dead town makes a troubling sight, reminiscent of cities ravaged by conflict. It turned out this ancient town had indeed been bombed, just not by man-made explosives. The Thera volcano produced 'volcanic bombs', also known as lava bombs: chunks of semi-molten rock thrown huge distances by the power of the eruption, ranging in size from a few inches to twenty feet across. On occasion they will explode from the pressure of their internal gases cooling, but even when they don't, they can cause shocking amounts of damage. We came to a wall which had been built of thick stone blocks. It must have been very strong. A volcanic bomb had gone straight through it like a strike missile. Akrotiri looked like it had been destroyed in battle: a war without any warriors.

We walked a few paces behind a tour group for a while, gleaning snippets of information from what their guide was saying. Akrotiri is known as the 'Greek Pompeii', but while Pompeii and Herculaneum are famous for the casts of their dead inhabitants made by pouring plaster into the hollows their bodies had made, no bodies at all were found here. This is because an earthquake had struck before the volcano exploded. The streets had been partially cleared of debris and all the residents had been evacuated, only – presumably – to die in the eruption that followed.

Having left the covered area, an eerie domain of dust and shadows, we returned to the sunshine outside and walked up on to a rough grassy bank. A few yards away, a small white cross poked out from a spot of stony ground whence you can look southward across the Sea of Crete. It was the grave of Marinatos, asleep in the place he had loved.

*

I had booked a dive with Santorini Dive Center, an operation based on Caldera Beach. Cand did not want to try diving again. She came with me to the beach and our dive leader, Marc, provided her with a sunlounger, a parasol, an unlimited supply of instant coffee, and the company of his cat. Marc was short and amiable with a close-shaven widow's peak of black hair and an unplaceable trans-European accent. He possessed what I now called 'the aura' – a subtle glow of competence which made everyone pay attention to his words. 'And Mata, my wife, will be always behind you,' he said. It was reassuring to know we would be followed on the dives by Mata, a divemaster who moved with the grace of someone who had spent months, as opposed to hours, underwater. On the surface she noticed everything and said little, as though she had exported from sea to shore the habit of living without having to verbalise every thought. She was present in her weathered black wetsuit at every dive briefing, sometimes carrying her baby daughter on her hip. Mother and daughter wore the same expression of contented calm. I caught Marc smiling at them as he geared up.

Our entry point was a rocky section of beach opposite the dive centre. Bits of pumice were drifting about on the sea – floating rocks, redolent of miracles. We walked into the water and within minutes had reached a depth of sixty-five feet. On a narrow shoulder of rock I spotted an amphora. An actual amphora, underwater, in the caldera of Santorini: a search engine's dream of Atlantis. My breath quickened. I swam over to it and took selfies. *You want me to get some pictures?* Marc signed. I floated beside the amphora while he took a few shots.

Then I saw what I had warned myself not to hope for. Ahead of us, un-dissolving out of the haze, there was a series of ridges that looked in every way as though it had once been man-made terracing, constructed with precision when this tract of seafloor had lain above the shoreline. A few small rocks had tumbled away to the side, as they would have from an abandoned structure on land. Even the way these 'terraces' were coated with a look of greater age than their surroundings, shaded green with a paste of algae, was reminiscent of the mossy coating on an old drystone wall: the proof of its antiquity.

As we glided towards the parallel lines of rock my imagination went to work like a digital special-effects program, superimposing the square houses preserved in the ruins of Akrotiri on to this underwater vista which had become, for me, the architectural ghost of the Mycenaean age. There were no ruins here but those a volcano had left, but give a human something that looks like a rampart, and on it they will build a castle of fantasy. It was wishful thinking. The 'Santorini–Atlantis' theory had got the better of me: the island, the sea, the Bronze Age, the destruction, the slope underwater . . . They had fused without my intention. I understood the appeal of fudging these things into a vision that smacked of Atlantis, in the absence of any sort of rational argument. *But so what if my Atlantis is fictional?* I thought. So was Plato's, and it's had a pretty good run.

Back ashore, I asked Marc about the amphora. 'Put there on purpose, for tourists,' he said. 'I'm sorry.'

★

For the next descent I partnered with an Italian technical diver and underwater photographer called Giorgio. He was friendly, grinning with happiness to be in Santorini. A cheerful buddy augurs a good experience. 'Apparently there is a little boat down there and I'd like to take some pictures, if you don't mind,' said Giorgio. Marc told us that now he had seen us dive, and we were familiar with the site, he was content to let us swim off and do our own thing. 'Just don't do deco, please. Or I'll have to wait with you,' he said. He was concerned lest we stay in the deep for too long and need to do decompression stops on the way back up, which would interfere with his schedule. Giorgio gave an enthusiastic OK, and I followed: we would mind our limits. This was a major milestone for me. I'd be left to my own devices with my buddy at a hundred feet, in open water. It was warm and still water, but all the same, I intended to prove that Marc's faith in me wasn't misplaced.

At sixty, sixty-five feet, a shift of mood occurs in the sea. Philippe Diolé described seascapes beneath this depth as 'countries without shadow'. In them, he wrote, 'Light comes from the very substance of the water, so that neither seaweeds, rocks, fish nor our own bodies produce shadows, not even over the sand of underwater beaches.' In the sapphire waters of the Caldera I saw that he was right. It was impossible to tell where the light was coming from. Everything was radiant.

As I waited for Giorgio to photograph the boat I began to feel happy. The elation intensified until I was convinced that all my problems had melted away, morphing into an exhilarating sense of intellectual clarity. I took out my dive slate and began to write down my thoughts.

A black hand appeared, waving in my face. It was Giorgio. He asked if I was OK. I shook my head in an attempt to communicate something like *No, man, I'm better than OK. I'm far out. I am saved. All is well.* I believed I had not only found Atlantis. I had solved the sea. Around my regulator, my lips wore an easy smile. Giorgio asked me what the problem was. *Problem?* I signalled. *No problem!* Then an OK with both hands, moving them back and forth. Confused, Giorgio looked at me for a second and suggested we rise in the water a bit. I cheerfully followed him up fifteen feet.

Suddenly I felt very different. Now it was over, it was clear to me that I had been experiencing gas narcosis. In the golden age of scuba, French divers referred to it as 'the rapture of the deep', but nobody used that phrase during our dive training, because narcosis can be dangerous. The problem had arisen partly because I had drifted down to 115 feet, a little deeper than I was certified to dive. Giorgio had saved me from the siren cocktail of depth, gas, carelessness and myself. I had played casual with a golden rule of diving. You must remember your priorities in order: *Dive first, situation second, communicate third.* I had maintained communication and awareness of my buddy and surroundings, but forgotten to monitor my own dive with sufficiently close attention. Warm, currentless seas can be as dangerous as rough ones. They lull you into a false sense of security, and they have lulled people to their deaths. I looked at what was written on my dive slate. It was a breathless expression of the same things I had been thinking about for ages: *The Atlantis myth is not history, but prophecy. A vision not of the past, but of things to come. We dream of our own destruction because it seems so obvious in a world so ill-suited to our existence: we see the great wave crashing in, because we can only half-believe we have lived here at all.*

It is possible that my serenity about exceeding a depth limit was itself a symptom of gas narcosis. I had entered the region of the water below a depth of a hundred feet, what Cousteau called 'the zone of the unforeseeable'. The name is frightening and accurate. At these depths, things – sea-creatures, problems – approach out of a deep darkness. The mental and physiological pressures affect every diver differently and in some cases there will be unpredictable outcomes.

After the dive I thanked Giorgio for keeping watch on me.

'I could see you was weird,' he said.

I recalled Gord's story of a dive when he had glanced down and spotted his son Dean behaving abnormally beneath him, looking at his hands like he had only just discovered them, as a baby might. Gord had reached down, grabbed Dean's BCD, and yanked him back up from narcosis, from the danger zone. It was Daedalus pulling his child back from the dark 'sun' of intoxication. I didn't have a father in the physical realm any more, but life kept reminding me that there was no shortage of caring people in the world.

The final dive of any trip is a melancholy moment. Towards the end, as we began our return to the light, we swam along while very steadily gaining altitude. Marc called this a 'swimming safety stop', because although we didn't stop, we were rising so slowly that it allowed plenty of time for any remaining gas to meander out of our flesh.

As we entered the shallows we glided gently over a meadow of *Posidonia*, the silver-green seagrass of the Mediterranean. It was waving lazily in the swell, and the wrasse patrolled above it like strange birds over the steppe. I noticed a breeze block and swam over for a look. One hole seemed to be half-full of small rocks. Suddenly they moved in that manner that is immediately indicative of life: not the tumble of gravity or the swirling about of water or wind, but a twitch, a motion that could only be caused by reflexes or sinews. I edged closer and peered into the hole. Looking back at me from behind the heap of stones was a single golden eye, divided by a dark brown lateral line. It glimmered, iridescent in the half-light. A little horn protruded above its brow, and the flesh around the eye was bumpy, pale orange and pink. It was an octopus. Below its eye, the broad opening of its siphon shivered subtly as it exhaled water. It drew the rocks in close to itself. Each of them was clasped by a sucker. They shifted in an awkwardly synchronised motion, forming a cowl of stone.

I gently pinched one of the stones and gave it a tug.

The octopus held on tight to the stone. There was something else besides muscular power in its grip. Its grasp communicated determination, and as I pulled a little harder, the octopus did too, mirroring the tug in what felt like simultaneous response. All the time its eye stayed trained on me, focused.

There may have been several reasons why this little tug-of-war felt so uncanny. In recent years, public appreciation of advanced invertebrate intelligence – in particular that of cephalopods including octopuses and cuttlefish, but also the inner life of crustaceans like lobsters and crabs – has rapidly grown, led by the work of marine biologists like the 'scuba-diving philosopher' Peter Godfrey-Smith. I had been reading about all this, and my mind had been changed. The marine biologist Alex Rogers had described staring at a lobster as 'somewhat like looking into the face of a fully armoured medieval knight', and this had made me realise how humans, with our elastic, muscular faces, are prejudiced when it comes to assessing the inner life of animals with exoskeletons. A total of thirty-four muscles enable the enormous range of human facial expressions, so our implicit belief is that feelings must be reflected in some tightening or rippling of the outer layer – the contorting of the skin. The lobster and the crab cannot pull faces. They are stone-featured, permanently veiled by a mask of inflexible chitin, and without even realising we are doing it, we interpret this as evidence of a lack of interiority. It is tempting to think of crustaceans as organic buildings, with a tough exterior, that, while clearly flickering with life, do not contain anything as grandiose as a soul. We cannot see how a portcullis-like face might be guarding an entire personality. We should know better. Conditions like Moebius syndrome, myasthenia gravis and Bell's palsy can all affect or disable facial expressions. There is something disturbing about the idea – rarely spoken, but often presumed – that eyes must be able to narrow, or mouths to form curves, or brows to undulate, in order for us to believe that a being feels.

Unlike vertebrates such as ourselves, whose neurons are largely concentrated in the brain, two thirds or more of an octopus's are located in its arms. The 'brain' is dispersed through the body in a manner far estranged from our own biology. Its arms contain interlinked yet independent 'control centres' that allow the limbs to act as semi-autonomous things: they make the animal, in a way, a community unto itself. The octopus wasn't just holding on to the stone with its powerful suckers. It was holding it using its mind, or a part of its mind. And yet its arms, if lost in violence, can grow back. It can lose its mind a little, and regain it.

I returned to myself: Giorgio and the other divers were getting ahead of me. It was time to let go. But just as I'd made the decision to do so, the octopus let go first, and my hand sprang clumsily back. It had won the game. And I hadn't even realised it was a game, until it was over.

I came here hoping to find, perhaps, an old Greek wall underwater, but instead I had met with intelligent life; had a tactile conversation with it. To look for man-made things at the risk of a missed heartbeat, or a twitch of life, was beginning to feel perverse.

But the interplay between the two things, life and waste, was a thing of its own. If a ruin is a place where nature has anchored itself to abandoned buildings, then what is a piece of rubbish inside the stomach of an albatross or a shark, or a speck of microplastic that has snuck behind the bacterium's curtain wall? There is such a thing as a ruin that's inside out: in which the abandoned man-made thing has colonised the crevice inside life, and not the other way around. We have no word for this 'ruin inverted'. Perhaps it is a thing too grotesque for words.

A month after we returned home from Santorini, the British government amended its Animal Welfare (Sentience) Bill to include marine invertebrates. The newspapers reported that due to research carried out by the London School of Economics into the central nervous systems of these animals, they would now be considered as sentient beings for the purposes of the law. A press release confirmed that the change would have no effect on existing industrial practices involving these animals, but that the new legal stance would be used to inform any future decision-making.

I thought of the octopuses hovering in the water at the time of the Thera eruption. They must have watched the silver surface darken at the wrong time of the day, perhaps darkening their own skins in response as they froze still, breathed the water, and observed. The octopus has also witnessed history and its destruction. When the volcano exploded, they must have fled as best they could from the source of sound, as above them the first streams of red magma were already cooling to black. Before long everything alive on the island would perish beneath a layer of hot white ash, but in the waters beside the deformed and dead islands of Thera, an octopus lived.

Palace of the Sea

Baia, Italy

Carlo beckoned once again and pointed at the floor: this time it meant pay very close attention. *Then, with gentle gestures of his gloved hands, he slowly began to brush the detritus away from a section of the floor. An area of small tiles was revealed. He pushed once again at the water and more sand rushed away. There were more tiles: dark ones alongside the light. A pattern was emerging.*

When we picture the drowning of human places, it is tempting to think it always happens fast. There must be an epic inundation, unfolding at speed. We can make sense of it as a happening of great significance. It is harder to fix our thoughts of tragedy and death on the slow up-creep of the water, the water that has no care for us and takes away our foundations at a pace of mere inches per decade.

In his poem 'Re-adjustment', C. S. Lewis uses the image of the drowning of Atlantis as an example of a dramatic final moment, which comes to symbolise, for him, the futility of mortal life if there is no God:

> I thought there would be a grave beauty, a sunset splendour
> In being the last of one's kind: a topmost moment as one watched
> The huge wave curving over Atlantis, the shrouded barge

Turning away with wounded Arthur, or Ilium burning.
Now I see that, all along, I was assuming a posterity
Of gentle hearts: someone, however distant in the depths of time,
Who could pick up our signal, who could understand a story. There won't be.

How much colder and devoid of meaning would the fall of Atlantis seem if there isn't even a huge and curving wave? A slow death is worse than sudden death. A fate with a sluggish, relentless gait is in its way more merciless than the smash of the rapid end.

★

There are few things on Earth more relentless than the volcanic forces that reshape our lands, roiling and flexing beneath our feet and under the sea. Many hydrothermal vents have been named with mythological breathlessness: Loki's Castle, Argus Tower, and the Lost City field, itself located — where else — on the Atlantis Massif, a marine 'mountain' that stands out from the Atlantic seabed as prominently as does Mount Rainier from the forests of Washington State. The difficulty of finding out who actually dubbed it the Atlantis Massif only adds to its mystique. This submerged island of rock, beyond the Pillars of Heracles, is where the poetic sensibilities of science take an odd turn, decreeing that the physical seabed should be labelled using ancient fictional nomenclature.

A map of the world's oceanic hot vents looks like an evil battle plan: chains of searing geysers situated in the most inaccessible locations on the planet, lurking deep in the oceanic darkness, where temperatures, pressures and the absence of light make for the most hellish conditions imaginable on Earth, from a human perspective. It is wondrous to think of the birth of stones when holding one in your hand. In a sense they are the fossilised remains of the life of Earth itself: skeletons of fire. Their solidity at 'human' temperatures — at every temperature which we can survive — indicates that it's always cold by the measure of their being: they are the ice form of rock. It seems absurd to think of temperatures that could melt the mountains like glaciers under sun. But such temperatures do exist, deep beneath us. Occasionally they re-emerge, and remind us we are only alive by the grace of our world's occasional predictability.

Life does not stop at the seafloor, even when it is made out of rock. It isn't just the rocks of the Pilbara and Lake Van that contain evidence of life within the stone. Microbes have been discovered living within oceanic rock both young and old. An impenetrable, dead environment turns out to be rich in life. Can we trust our poetic instincts on anything? Why obsess over finding Atlantis when the rock itself might be alive?

We call the creatures that live in such places 'extremophiles' — lovers of extraordinarily harsh environments. But of course, their circumstances are normal to them. From another's perspective, we humans are extremophiles of a different kind, sipping drinks from vessels of frozen sand and dying of heat exhaustion at temperatures that would be fatally cold to thermal vent crabs. In turn, we have created our own extremes that everything alive must now cope with. Indigestible fabrics, oceanic dead zones and a hotter planet mean every living thing might need to find its extremophile streak to stay alive.

★

I asked John, a close friend since university days, if he wanted to come with me to Naples and see the ruins of Baia. John agreed and I was glad to have a buddy who would follow me into the water. Freshly parched from the filtered air of the plane to Naples Airport, John and I boarded a bus from the terminal to the centre of town. Through the windscreen I saw a grand avenue ascending towards the light of the sky like a ramp directed to heaven: it was the Via Casanova, and at the end of it was a high place, golden and misty as if set within a haze of sparse cloud. This was the Castel Sant'Elmo, the old fortress on the hilltop of Vomero. In the afternoon sun it had the look of a seat of gods.

We got off the bus and descended into the underworld. The metro station was a place of dark metal walls and staircases sinking to nowhere. It was deserted. With no one to follow, we took a wrong turn in the labyrinth until in the amber half-light we noticed a frightening shape lumbering towards us. In the gloom it appeared to be a figure with broad shoulders, three heads and six legs. As it came near, its central head shouted at us: '*No! Fermare!* – No! Stop!' This Cerberus turned out to be a ticket inspector flanked by two young women who he seemed to be escorting through the caverns. He barked at us for our tickets. We showed them and he sent us the other way, then shuffled back into the dark.

At the entrance to our apartment block we met Branko, a stoical ringlet-haired thirty-year-old with a calm smile framed by the lazy trace of a beard. We were late but Branko didn't care. He ushered us through the small door in a great wooden gate, the kind of entrance that in Britain means 'ancient, elite university', and in continental Europe means 'block of flats'. He showed us the lift – a dry-cleaning cabinet hung from a black cord – and the stairs. John and I looked at each other. We made for the stairs.

In our flat there was a mixture of beautiful original furniture and flippant art. A triptych painting over the dining table showed an exploding Mount Vesuvius, the pyroclasm represented by squiggles of red and black paint. The mountain and foreground were done in thick white strokes. Black, white, red. These were the colours of the beaches of Santorini: the volcanic *tricolore*.

The roads of much of western Naples were barred for a cycling race, the Giro d'Italia. Unable to make it to Baia by road, we needed to use the trains. Nobody we spoke to had heard of Baia. It was only fifteen miles from the city centre but, perhaps because it was now underwater, or because the Campania region is full of ancient monuments, it had ceased to occupy locals' minds. Our bad Italian was not helpful. John made a cocktail of Spanish and Latin at which a rail worker raised one, then both, eyebrows. But it worked.

We were directed to Fusaro Station and arrived with exactly fifteen minutes to make the fifteen-minute walk to the dive shop. Terrified that we might be late, which in England would mean the dive boat had cast off ropes and gone without you, we rushed through the heat and arrived at the shop with a minute to spare. Inside, the divemasters looked serene. They smiled at us. A stocky shaven-headed man with a beaming smile stepped forward.

'Good! You are here. I am Carlo.'

'I'm so sorry we didn't get here sooner. The roads are all –'

'I know! No rush! You can relax. We figure out some logistics.'

'Is it OK if we eat our sandwiches?'

'Of course, my friends! Sit! Eat!'

There was no sense of time being critical.

'This is not like home,' I said to John.

The divemasters dimmed the lights of the room and turned on a projector. Its blue lamp cast the interior of the shop in a gloomy shade of the sea. They took it in turns to present slides about the history of Baia. We were told about the volcanic volatility of the area: the nearby Campi Flegrei, the Phlegraean Fields, is one of only twenty known supervolcanoes on the planet. In spite of the many well-known eruptions at other locations in the area, most infamously the explosion of Vesuvius that destroyed the towns of Herculaneum and Pompeii in AD 79, the last great discharging of the Phlegraean supervolcano had occurred 39,000 years ago and it had 'mostly' been quiet since. *Mostly*. This observation was meant to quell our fears about standing on top of the mythological home of the Roman fire-god Vulcan, counterpart of the Greek Hephaestus. All I could think about was how thin the crust of the Earth was here. It was surprising that Baia had been engulfed by water: lava would have been more fitting.

The dive site is famous for its statues. These stood in the emperor Claudius's Nymphaeum – 'the room of nymphs', sculptures of heroes and royalty. They were replicas, 1:1 scale copies of the originals which had been removed for their own safety to the archaeological museum in the nearby castle of Baia overlooking the bay. John and I looked at each other and gave a simultaneous shrug of resignation. *Copies means fakes.* I already knew what had been done with the statues. The need to preserve the original sculptures from suffering further damage was clear, and the replicas had themselves been in the sea for decades now and probably, as a consequence, looked just as sea-kissed and enchanted as the originals once had. I still didn't like to be reminded of it. Having travelled a thousand miles to get here, we would be staring at imitations. It wouldn't be quite the same as looking into those marble eyes that had kept their 2,000-year vigils, first in the pre-industrial air of the Roman Empire, and then in the changing light of the sea that has never acknowledged the hegemonies of men.

'Please, don't take souvenirs. It is an archaeological site after all,' said Carlo, who would be our guide for the day. 'Most of the ruins are sitting now in fifteen feet of water. For diving it's not a lot, but for the villa, it's maybe too much.

'You should not have to worry about your air. Obviously we are in quite a shallow site, so if you manage to get down to fifty bar then well done, we promise to buy you a beer. But please, don't do it just to get the beer.'

Enrico, a younger divemaster who had been standing to attention at one side, interjected to discuss the phenomenon of *bradisismo*, or bradyseism: the slow, upward or downward motion of the ground in an area which is known for volcanic activity. As underground chambers empty or fill with magma or water, there can be great expansions and contractions of such spaces, causing areas of land to move up or down. This has happened in a pronounced way at nearby Solfatara – from the Latin *sulpha terra*, 'sulphur land' – a volcanic crater in the commune of Pozzuoli. Even here, where we sat, the land was rising by four centimetres per year, slightly faster than the rate at which the continents drift apart. As Enrico explained these things I noticed that there was a difference between these divers and those who worked in the crater of Santorini: here, the signs of current volcanic activity were everywhere. The land was pocked with sulphurous sores: there was hot steam rising from beneath. As a result, our dive briefing was full of references not just to past eruptions, but to the places where the heat of the Earth was still sizzling to the surface. Santorini, in spite of its being half-covered with a thick layer of pumice and ash, was a place where even small eruptions were the stuff of legend, not something you might notice signs of on the commute to work. Here, the magma roiled away just two miles beneath the streets. That sounds like a long way down, and in a sense of course it is, until you think that the average cruising height for a commercial airliner is between six and eight miles above the Earth.

Carlo pointed to a rectangular area on a map of the underwater site.

'Here you will see the pool. Still full of water! Ha ha!' he said. The other dive guides smirked and shook their heads.

Ultimately Baia had been a resort: a place the wealthy visited to relax, let off steam, or engage in a little Roman debauchery. 'People were having fun here, for sure,' said Carlo. The town has been referred to online as a Roman Las Vegas.

'You remember your signals? OK? It's both a question and answer,' said Carlo. 'Well, for me it's a question, for you it's an answer. Because nobody cares if I am OK.'

Lastly, Enrico warned us that the viz might not be great – we were to expect anything from six to twenty feet – and he explained that when we descended, the mosaics would all be hidden under sand. They were deliberately kept like this in order to prevent algal growth. We would swim up close to them, and then Carlo would brush away the sand to reveal the mosaics.

'Doesn't that damage them?' asked John.

'No. And remember, they survived fine for some thousands of years before we started to even care about them. They can survive a little brushing of the sand.'

There did seem to be a contrast, though, between the attitude to the mosaics and to the statues which had been removed to the museum. I reasoned that it must have been because statues are more likely to be damaged by boats. And they would be easier, and perhaps more profitable, to steal.

We left the shop and walked around to the small harbour behind it. Once on the boat I readied my gear and prepared to dive. The water was shallower than John and I had expected, and the visibility good enough that the ruins would be visible from the surface. John opted to snorkel above the ruins with Enrico, viewing them from a 'bird's-eye' position. I descended on scuba with Carlo in the hope of getting some close-up photographs of the site. By this point in my fledgling life as a diver I had realised that although the best pictures tended to be in my head, the problem was that only I could see them.

I got the signal from Carlo, held my mask and mouthpiece in place with my palm and backrolled off the boat. The thrill of entry having passed, I steadied myself and began to look around. It was as Enrico had warned: the sea was a soft, milky blue, the shade of a curaçao cocktail. Through it I could see perhaps ten or fifteen feet. That would have to do.

Carlo and I gave each other the OK, and he started to swim off with purpose. I followed. Carlo moved very slowly, like a bomb disposal expert for whom false moves weren't an option.

He summoned me to come closer. We had neared the outside corner of a ruined building. It was in good shape, still forming a sharp L of stone, several rows of bricks high, and crowned with a dusty blonde hairstyle made from a fuzz of seaweed and hydroids. A couple of small bream joined us, expressing their curiosity with the puppyish body language I had come to expect from fish acclimatised to the company of divers. They fell in behind us, among the dust which our fin-strokes wafted up from the floor, and jetted about, picking off interesting bits of matter: some they spat out, others they swallowed. We were hosting a tasting party in our wake, but our attention would soon be fixed on the ruins, the scale of which was beginning to be revealed.

Carlo moved towards a gap in the wall and shone his torch on a threshold — the remains of a doorway, marked by a rectangular slab of purple marble. He brushed away some sand that had collected on top of it. The marble was in fine condition, almost surreally so: its edges were sharply delineated and its face was flat and smooth, with only a couple of small chips to speak of the fact it had spent 1,500 years underwater, and prior to that, been stepped upon by countless human feet. I was staggered to see this piece of flooring in such a good state of repair. As far as I knew, these parts of the ruins were entirely original: it was only the statues which had been replaced with replicas. I couldn't quite believe it. A piece of marble left outside in England for even a fraction of this time — a gravestone, for instance – would have been savaged by the elements, clawed at by wind and frost and baked brittle by the sun. The sea had been this marble's formaldehyde.

Carlo beckoned me over to what looked like a staggered section of wall and pointed out a row of holes in the top. He took his alternate regulator and put it into a gap underneath the front section of the wall, then pressed it to let air out. The air flowed up through the wall and columns of bubbles rose, uniform and silver, from the exposed holes in the top. It was the thermal ventilation system, and it still worked. Carlo put his regulator away and used a mixture of standard scuba hand signals and tourist-style improvised gestures to say more. *The hot air would circulate under the floor of the building and up through the walls, so you would not be cold here. OK?* He spoke fluent 'diver'.

Between the marble, the walls and the ventilation system I had already been wowed, but there was something about the way Carlo beckoned me to the next location that made it clear it was going to be special. He gently floated down to what had been the corner of a large room, and gestured to pay attention to the floor beneath him. He began to push rhythmically at the water, creating a wave which swept sand and bits of shell and organic matter off what lay beneath. It was one of the mosaics. The tiles were a dirty-white colour. Carlo swept away some more and a section of black tiles was revealed.

Is it a leg? I thought. It was. Carlo continued sweeping, sending careful pulses of water into the sea-dust and blowing it off. Eventually he was finished. He moved his arms over the results as though unveiling a banquet, inviting me to take it all in. It was the mosaic known as 'The Wrestlers', depicting two black figures clasping each other's arms. The bodies were simplified, but well observed, full of tension and focus. The definition of their muscles was shown using lines of white tiles. The one on the left was grimacing, his bicep bulging under the stress, while his opponent seemed more relaxed, as though untroubled by their underwater combat, in which they had been locked since before the fall of Rome.

This mosaic was Roman, but it was the most Atlantean thing I had ever seen. The schematic faces of the men and the mystery of their ethnicity played into it: this was the art of a lost, warlike empire, hidden on the seabed, preserved under a shoal of mud. The bream and rainbow wrasse moved in to pick off particles of organic matter disturbed by Carlo's sweeping, reminding us where we were, on the bottom of the Tyrrhenian Sea. The detritus that lay around in heaped piles, neatly swept aside, made the mosaic look like it had just been dug up, revealed to the light and to human eyes for the first time in 2,000 years. *It must be strange for Carlo*, I thought, *unmasking this every day, re-enacting the miracle of discovering the same piece of ancient art over and over again.* But when I looked at him he was hovering in a kneeling posture, levitating, hands clasped in his lap like a monk of the ocean. He was silently contemplating the mosaic as if praying before an icon. It was obvious that he didn't find it boring. He looked like he was experiencing the latest instalment of an iterative epiphany. I returned my gaze to the mosaic to take it in for the last time before he covered it back up, and as the heads of the figures disappeared again under the dark marine soil, I shook my head in disbelief to have replaced my dreams of a place like this with the sight of the real thing.

Reluctantly, I followed Carlo as he swam to another section of the town. On the way there he paused and picked something up. It was a dull shade of orange. To my surprise he turned and handed it to me. It was the handle of a terracotta pot. My heart skipped with sudden nerves. Was I allowed to touch this? Evidently yes, or Carlo wouldn't have offered it. I turned it over in my hands. With his index fingers Carlo traced through the water the outline of an amphora. Then he pointed at the seabed. *That's from here*, it meant. Someone had gripped this handle, long ago, when its vessel was full of a valuable cargo. Oil, barley, vinegar, wine. Grasping it with my own hand made the whole experience of diving in Baia more real. Seeing was one thing. Touching was another. *I am here*, I thought. *It's in my hand.*

I gently placed the pot handle down and followed Carlo, who began to uncover a second mosaic. This one was not figurative, but bore a pattern of four large decorated circles with a square in between. It is known as 'The Shield Mosaic', because the circles resemble round shields of the Greek hoplite kind. It was just as stunningly preserved as 'The Wrestlers' and, as a timeless design which might be found in a contemporary house, it was, in a way, more eerie. I could imagine this pattern on the floor of my home. This was someone's home once, before it was repossessed.

Our last stop was the Nymphaeum of Claudius. By now the current had dropped away and I was able to drift slowly, silently. The room was long, with a semicircular apsis at the far end, and the floor was piled high in places with the underwater 'soil' of Baia, giving the impression that this had been a ruin above ground long before the water took over.

I needn't have worried about being disappointed by the replica statues. This was one of those occasions when the relatively 'poor' visibility only served to heighten the intensity of the experience. I drifted through the misty chamber until without warning, out of the teal-coloured light, the statue of Dionysus un-dissolved itself. It appeared softly but suddenly, as a ghost might. A gentle kick of the fins sent me further along the room. There was the statue of Antonia Minor, mother of the Roman emperor Claudius, and the two headless figures of Ulysses, one holding a bowl in his arms, the other down on one knee and offering a cup of wine now full of seawater.

The statues were successful replicas. They invoked a numinous atmosphere, of awe before human and natural achievement. It was a homage to creation, exalting the art precisely as it spoke of its susceptibility to decay. Not only were these statues a perfect example of Atlantean classical beauty, they resembled Plato's Atlantis story at another level, by drawing those who swam towards them into their fantasy.

★

In the wake of diving in the ruins my eloquence felt blunt, compromised. In the Spanish quarter I unwittingly offended a young bride-to-be who was dressed, bizarrely, as a red leather handbag. That evening, John and I went for dinner at a small family restaurant serving Neapolitan onion stew. It had simmered for days so that we could devour it in seconds, and I struggled to clumsily phrase my appreciation. My mind, more so than after any other dive to date, had left some part of itself in the water. I wanted to be back among the nymphs.

On the flight home I gazed sadly at the sea below, where yachts from that distance were indistinguishable from ancient barques; I pictured them manned by brave sailors, bare-chested and bronze, cutting straight paths across the peaceful Tyrrhenian, its spectrum of blues in impossibly smooth transition.

From the sky, the great terraces of Naples were clear to see. It is a city staggered upwards from the shoreline, as if built in readiness for its tiers to be abandoned one by one should the ocean rise. The effect was strengthened by the visible lines of depth in the water as it deepened away from the beach, charting a reminder of the land it had taken before. People say the sea 'swallows' the land but it doesn't: the sea incorporates. The land that slips under the water becomes a part of the sea's reality, part of its realm. From such a height the unintentional comedy of any attempt to contain the sea, any hope for eternity on Earth or stability in our lives, becomes clear. We have built frail crusts along the edge of the water, and have no more guarantee than the barnacles that conditions will remain favourable for our life.

My chest was gripped by the cold feeling of a sudden crisis, and I was shocked to find I was crying. In our keenest moments of grief or love, salt water emerges from the secret sea inside ourselves.

Dad adored salt. It is possible that his love for it contributed to his death: high sodium consumption has been linked to the onset of strokes. The taste of salt reminds us of our origins in the sea. Too much of it prompts us to remember we have changed. We no longer live in a salt-world. We cry tears, but we are not supposed to dwell in them.

Tears, besides water and salt, contain a mixture of lipids, lysozyme, lipocalin and glucose. They are an antibacterial liquid. They flow when something needs to be washed away: in this case, it was my delusion that I was more in love with ideas than I was with the world.

'Seek not to speak soothingly to me of death, glorious Odysseus,' said Achilles from the shadowy afterlife of Hades. 'I should choose, so I might live on Earth, to serve as the hireling of another, of some portionless man whose livelihood was but small, rather than to be lord over all the dead that have perished.' Better a serf on Earth than a king of shades.

In the Beast

SS Fortuna, *English Channel*

I swept my torch beam around me. In clearer waters the bulb would have shot out a post of light, but in this cloudy blackness it could only produce a stubby fan, a floret of yellow that barely reached a yard beyond me. These waters had a reputation and I was learning why the hard way. A person could lose their mind down here. Just hold it together. Stay calm.

I was back in England and the chance had come up to do a wreck-dive in the Channel. I opted in but I was nervous. I hadn't dived far offshore in British waters. They can be challenging even in the clement months of the year, and are graveyards for several thousand years' worth of boats. Maps of the shipwrecks in the English Channel are shocking to look at. On some of them, the crosses and dots that mark the wrecks are so dense they form blocks of colour, with only tiny patches where no vessels are known to have sunk, little plots of relief in a slew of disasters. By some estimates there are 37,000 shipwrecks in this treacherous sleeve of water, many of which have never been seen by human eyes since they sank, and that's only the ones that are accounted for. The boats of smugglers and unwatched sailors are down there among them, unlogged.

We were due to dive the wreck of the SS *City of Waterford*, which sank in 1949. The *Waterford* was a 1,334-ton steamer that collided with a much larger Greek ship en route from Antwerp to southern Ireland. Prior to that she had navigated the war and shot down a Nazi plane while carrying livestock across the Irish Sea. They must have whooped and cheered on deck. The captain had been decorated for his part in the action. Did they guess their ship might one day join the plane?

There were five of us from Ocean View aboard, and three technical divers – one of them with a closed-circuit rebreather – who I didn't recognise. Once all the divers were on the boat, the skipper, Steve, came out to address us.

'Right. We've had some strong current coming up from the south. It's blown all the particulate off the bank in the direction of the *Waterford*. We had divers on the wreck earlier today and they were reporting viz of six or seven feet. If that.'

A few heads dipped.

'Look, I'm happy to take you there if you're willing to accept the risk with the viz. Or we can go for a wreck that's further from the bank and hope for better conditions. I was thinking about the *Fortuna*.'

'*Fortuna*'s a nice dive,' said one of the technical divers.

'Shall we go for that?' said another voice.

'Yeah, I don't fancy a pea-souper,' said someone else.

'Is everyone OK with that, then? We'll try the *Fortuna*?' asked the skipper.

There were no objections. The *Fortuna* it was: Latin for 'luck'. I hoped our luck on the dive would be better than that of the ship.

We motored out into the Channel, which was choppier than any of us had expected. The little boat bucked and dipped and we held on tight. The fact that I was a novice diving with men and women who'd spent many an hour in these places heightened the sense of ill ease. I felt like an innocent citizen who had stumbled into a town where the people were privy to the secrets of an underworld. I would have to survive it for myself, as they had.

I then did something that in retrospect may have been a bad idea. I looked up 'Wreck of the *Fortuna*' on my phone. It was a 1,254-ton Dutch steamer that had sunk on 22 October 1916. The *Fortuna* had struck a German sea-mine. Fifteen of her crew had gone down with the ship.

Perhaps men had gone down with the *Waterford*. I didn't know. And that was the important thing: that I didn't know. Now I knew that the *Fortuna* was not just a shipwreck, but a tomb. A gust hit me off the water, a ghostly cuff. Maybe some divers wouldn't be affected by the knowledge that a ship had gone down with hands, but for me there was no way it would feel the same as diving on a wreck that I knew everyone had been rescued from. The *Fortuna* had been a death-trap at least once before.

We made it out to the dive site and kitted up. Phil, who'd come aboard for the ride, took a photo of me in my gear, sitting on the gunwale bench. I smiled and held five fingers up to represent that this would be my fiftieth dive. It had taken me two years to get here, and many frustrating waits. In different circumstances – less volatile seas, no lockdowns – a novice could log this many dives in a matter of weeks. So I felt proud of my stubborn progress, but for the briefest instant I wondered whether acknowledging my 'fifty' was a stupid thing to do. The dive hadn't happened yet. Maybe I should've waited until we were back aboard, and I actually had fifty dives under my belt, to make any kind of claim on the number, humble as it was. I tried to prevent myself from thinking the words *tempting fate*. Sailors, and others who work on and in the sea, a place unfriendly to human life, have traditionally been superstitious. The tendency was rubbing off on me.

By now everyone else had dropped and the skipper called out for the last pair: Stuart and me.

We checked each other's tanks were on one last time, then walked to the stern and made ready to jump. I had a final glance over my gear, then turned to my right and made eye contact with Stuart. We gave each other a nod. I looked out at the sea. It had a resemblance to blue obsidian, chipped and glum. There was no foam. Bit of a surface current maybe, but not bad. The buoy was eight or ten yards away. Not far. Beyond it was the coast, and home. That did look far.

I pulled my mask down and tried not to think about it. We waited for the horn.

Suddenly there was a call from behind us. 'Hang on a minute. Someone's coming back.'.

'What?' I said.

'Sit back down, guys. Someone's got a problem.'

It was one of the three tech divers. He had resurfaced and was swimming to the boat. He tapped his dive computer: something wrong with it.

Feeling the weight of our gear by now, Stuart and I returned to the bench. We looked at each other and each gave a little shake of the head. It didn't feel good sitting back down after being on the cusp, ready to jump, but it was out of our hands. Somebody needed help. It could have been us.

The tech diver got on to the tail lift and hauled himself back aboard. 'It's just not firing up,' he said. 'I don't suppose you've got a spare one kicking around?'

I would not have expected the answer to be yes, but Caroline, the first mate, went into the wheelhouse and came out seconds later with an old puck-style computer the size of a half-tin of beans. She strapped it on to the tech diver's wrist. It started up straight away.

'Lucky me,' he said.

'Lucky you,' said Caroline. She looked him in the eye.

'These things are tanks. Never seem to let you down,' said the techie. 'Thanks.'

He stepped off the boat to rejoin his buddies. They were in the dark, probably waiting for him at the bottom of the line, and I admired how casual he looked about it all. I had never descended alone. The thought of doing so miles offshore made me swallow my nerves.

Once we had the OK from the skipper, Stuart and I nodded at each other, got in position, and strode off the stern.

We made straight for the buoy, took our bearings, and descended. The viz was not great but it wasn't the worst we'd had. We felt our way down the shotline. My ears weren't equalising as readily as usual and I rose upwards a couple of feet to ease the pressure. Eventually I was equalised. I gave Stuart an OK and we continued downwards.

The water was darkening. The viz was not going to be good. *Never mind*, I thought to myself. *We're on our way now. Here goes.*

Then I felt something. My BCD tugged at my torso.

I was caught on the line.

No problem, I thought. *It's just a clip.* I moved my left hand up to undo it. It was awkward. The shotline was thick and it filled the hole in the shackle. I really was caught.

Entanglement.

It was Rán. My relationship with the sea had changed, but she hadn't changed her ways. She would still ensnare the careless or the unfortunate.

I thrust Rán out of my mind. There was no room for superstition. I reached out and tapped Stuart's shoulder. He turned and came in closer. I pointed at the clip and gave the horizontal wave of the hand that means 'problem'.

Immediately Stuart got to work. *It'll take him a second*, I thought. I worried a little when it started to take him some time. He put his torch in a pocket so he could engage both hands. *What if he can't get me off? What are we going to do? Shall we cut the line?*

I felt a small pluck and the tension dropped. Stuart had managed to get the clip off. I was free.

I continued moving slowly down the rope. It felt like a bad omen but I dismissed that thought as silly. It was a minor hiccup. We would get down there and, if nothing else, at least see and touch the ship's corpse. Is it fair to call it a corpse? The eels are in there, maggots of an iron cadaver.

I was puffing a bit from the stress of having become entangled while trying to equalise my ears. It was getting dark. Streams of particulate rushed through the beams of our torches, rivers of dust and tiny life forms flowing in wide, reflective sashes through the water. As something recollected – as a memory kept at the safe distance of time, at 'harm's length' – it might be beautiful, but in the moment it was foul weather. The particles were further complications in a hard moment, like a snowstorm descending on a midnight walk.

By the time we neared our target depth of eighty-four feet it was almost pitch-dark. The thought that it was a summer's day back at the surface was absurd. A murky blizzard swirled around us. We had moved through a portal; been catapulted from the season and out of adjustment. As we hauled ourselves along the rope it felt less like a journey downwards through water, more like a tugging through consequence.

We were approaching eighty feet: the wreck had to be very close. Suddenly Stuart was in the midst of my torch beam, an inky human shape with the shadows of a few wires plumbed into its outline, a man on a drip. Reassuring as it was to see him, he cast an alarming silhouette. Then my fins hit something vertical and I knew we must have finally reached the *Fortuna*. I looked to the side to make sure I could still see Stuart. I could make out the swoop of his torch. A glimmer of reflected light picked out his mask like a strange black skull in the dark. I resolved to check where he was every few seconds. I didn't trust this place. It felt like it might seize you away without warning.

I reached a gloved hand down and grabbed at the thing my foot had struck. A thick sheet of metal: the hull. It was strong and unforgiving and through the neoprene of the glove I could feel the unevenness of the rusted iron, gnarled like bark. I hadn't planned to touch the ship, but after the frantic journey down I was yearning for the reassurance of something solid. The ship had been lost but for me it was a port in a storm. I breathed out with relief, discharging a huge jumble of bubbles. My computer gave me a depth of seventy-nine feet. It was only a nick in the abyss of the ocean. These paltry-looking numbers mock the experience, the overwhelming sense of being in another world, as far from safety as I had ever been.

I took a moment to steady myself, check my gear and shine a torch on my compass, depth gauge and watch to make them glow green against the darkness. I trust my computer but there is something reassuring about these mechanical instruments which operate without batteries or what, to me, is incomprehensible circuitry. They are full of simple springs and joints and valves responding to pressures, encased so they can operate underwater. They are a little bit more like me than a circuit board. Unlike the dive computer I can understand and appreciate how they work. As long as these gauges keep on ticking and measuring, I will keep ticking as well. My heart will beat and I will be OK.

The first known analogue 'computer' was itself more like a

mechanical watch than any sort of cyber-device. It was the Antikythera mechanism, salvaged from a shipwreck off the Greek island of the same name in 1901. By then it had spent over 2,000 years in the sea, and a corroded blue-green lump of metal was all that remained of what had once been the most complex calculating device known from the ancient world. Twenty-first-century X-ray analysis enabled researchers at Cardiff University to take photographs of 'slices' of the artefact. These showed that it had once been a clockwork orrery composed of thirty-seven intricately fashioned bronze gears, able to track the motion of the sun and moon and account for the irregularities of the lunar orbit. It may also have been able to predict eclipses and the positions of the five planets known to classical astronomers – Mercury, Mars, Venus, Saturn and Jupiter. It is possible that Hipparchus of Rhodes, who played a central role in the invention of trigonometry, was involved in its making. Reconstructed versions of the mechanism, using glass or Perspex in place of its original wooden case, enabling its workings to be seen, are beautiful objects: curious clocks without faces, driven by a handle like the crank of an old motor car. It would be a millennium and a half before humanity redeveloped the know-how to build similar devices, in the form of the *horologia* or astronomical clocks of the late Middle Ages.

We take it for granted nowadays that a timepiece will run like clockwork, but an astonishing amount of engineering and testing goes into the making of a diver's watch. Its crown, the little fat disc you turn in order to wind it and set the hands, screws down against the main body of the watch to keep water out. This is not simply a case of a threaded screw. The crown is actually a multipartite spring-loaded door with one or more tiny gaskets inside, as complex as a submarine's entry hatch, only thousands of times smaller. This is the sort of intricate work that prevents the sea interfering with your schedule and calculations, that keeps time dry. Proper diver's watches weren't invented until the early 1950s. A couple of the most famous models first came out in 1953, the same year that the double helix structure of DNA was discovered. It had taken humans as long to supply divers with waterproof clocks as it had done to model the building blocks of life.

What a mad idea, in any event, that someone needs to wear a wristwatch at the bottom of the sea. It is like needing your car keys at the top of Mount Everest, or your business cards while jumping out of a plane. Nowadays there were glitzier ways to keep track of the time underwater, but a clockwork device held a charm that bestrode the ages, the 2,200 years between the cogs in my watch and the making of the Antikythera mechanism, itself built when the story of Atlantis was still young. But only now did I realise the most important function of the diver's watch, the one which goes unmentioned. It was to trick you into believing that the place you were in was normal. *I am wearing a watch. It can't be that different to another day at the office, or a walk in the park.* It frees a part of your mind – the part that should be constantly freaking out – to think about something else. No wonder divers are so often attached to their watches, the only part of their aquatic gear that they don't take off when re-entering life on land. The watch is a talisman that links the wearer's home world to the sea and which, like a Roman soldier's armilla, symbolises their bravery. It is only a watch, and that is precisely why, in tricky moments, it is also much, much more than a watch.

The water was rushing and swirling and in the scattered light of our torch beams countless tiny creatures and particles of organic matter whirled and churned, lit up by the harsh LED bulbs and shining bright white. It looked like we were swimming through the deeps of the universe, a tumultuous night sky. When you gaze up at the stars they move so slowly that, beside the comets, their motion is imperceptible, but for a being that lived longer and saw differently, the spinning of the heavens would be like a whirlpool, a celestial zoetrope. This is what the waters around the *Fortuna* looked like to me. Perhaps this is what the unfurling of the cosmos would look like to a lesser god, diving through the aureoles of time. It should have been beautiful but I felt like I was in a nightmare from science fiction. Alone in a churning shadow world.

Alone. I looked to my right for Stuart. Nothing. I must have waited a few seconds too long to check where he was. I swung my torch-hand out towards the direction I had last seen him in. Suddenly the hazy beam bent as though directed by a mysterious force. Then my hand hit a wall of metal. I dropped the torch and it hung from the safety loop which, luckily, I had remembered to cinch around my wrist. My God. I must have accidentally ended up inside the wreck. Fear rushed up through me. This was not meant to happen.

The thought of my family stung me like a shot of adrenaline. I felt ashamed of what my death, if it happened, would mean to the people I loved. My death would be stupid. I would not have been down here trying to rescue someone, or mend an oil rig or communication line. I'm not even looking for actual treasure. What if I die like a fool in search of Atlantis?

Was I panicking? I feared panic the most, more than sharks or entanglement or even running out of air: when a scuba diver dies, more often than not it is panic that kills them. Becoming irrational underwater is often a shortcut to death. People pull out their regulators, take off their masks and fail to ditch their weights. The panic manifest in deep and rapid breathing. As if your soul is sprinting even though your body is still. It is two selves fighting: your animal self, screaming, *Get out of the sea now*; and your diver's mind restraining, wrestling it back down. I checked over my instruments to make sure my mind was still sequencing things. I hadn't done anything stupid. I didn't think I was panicking. But I was definitely very scared.

The shock that I had inadvertently drifted into a shipwreck caused me to take in several deep draughts of air, and with the increase in buoyancy I floated upwards a few feet. Suddenly, there was a torch and above it the huge, dark eyes of Stuart's mask. I must have slipped just below the top of the hull wall, which had been screening Stuart from me. The relief was immense. And there was something else. Behind him I could see a tiny trio of lights, bright specks in the blackness like three stars in a night sky. It must be Paul, Teresa and Mark. Stuart and I weren't alone.

I looked at my watch. Five minutes: a mere five minutes had passed. It was further proof that the sea had its own time. You can take a waterproof clock with you, but the system of time measurement that it displays was conceived of on land: in the water it transforms. The second hand that usually seems to glide so smoothly appears, when magnified by submersion, to be jittery and unsure of itself, as though nervous about being in the water. It wasn't that this five minutes had felt like hours, or raced by in a flash. It was more like it had dissolved, represented by the swirl of numberless motes of dust in the water around me.

In the corner of my eye I thought I saw the lissom S-shape of a conger eel, silvering its way through the darkness, but it turned out to be just a ribbon of particles caught in the edge of my torch beam. I checked for Stuart again then looked at my computer. Its display had turned yellow and a time alert was flashing: 2 MINUTES REMAINING. What the hell was this? We had only just got here. It couldn't be a deco message. Then I saw another phrase: AIR WARNING. My eyes widened as I recognised that I had been motoring through my air. The issue with the shotline must have stressed me. I had used almost half my cylinder in only ten minutes. I thought about the length of the journey back to the surface, and the fact that we needed to stop for five minutes on the way up. Was I going to have enough air for this? I struggled to do the sums: air consumption, depth and ascent time. My mind conjured only colours and temperatures, whiteness and cold. I couldn't do the maths. I was losing control. Then I looked to my right and saw that I had lost Stuart for the second time. Nor could I see any of the other divers.

I realised just how out of touch with the surface we were. We had been instructed to pull on the shotline to jiggle the buoy if something went wrong, but I had lost the line. There was no radioing the captain from deep underwater. We were fending for ourselves.

My breathing had started to quicken again when Stuart re-appeared beside me. I knew it was probably time to end the dive. I gave Stuart the 'thumbs-up' signal and he copied it back.

Often while diving I think about the holographic layer of control you must create: superimposing calm over your instinct, which remembers it ought not to be underwater. Back on deck, I realised that a part of me had actually lost control: the deeper part, the part that felt more real. Somehow the trained, concocted part had managed to gag it and tie it down. I looked at the dive profile on my computer. We had kept to a normal ascent rate. We had executed a free safety stop very well, hovering in moving water. The only thing that looked weird was the brevity of the time we had spent at the bottom. You could never infer from this graph the stress I had experienced. The training had worked. I was disappointed to have seen so little, and more so that I had shortened Stuart's dive. But in among the discontent was a sliver of satisfaction that I had managed to behave as I had been trained to do, and that I had grasped the ship, and was still alive.

Teresa climbed back aboard.

'How was that?' Phil asked her.

'Slightly terrifying,' she said.

Mark and Paul emerged in turn at the top of the ladder. They were grinning. Pitch-black shipwrecks, crevices, eels and tricky navigation were their sort of thing. *They're made of something different to me*, I thought.

The three tech divers came out last. The guy in the closed-circuit rebreather made his way over to where we were sitting. He'd probably done a thousand dives. Surely he'd had the time of his life down there.

'Well. That was tough,' he said.

It was a surprise to hear it from him, and not an unwelcome one.

'Tough viz,' he added.

His buddy, who'd dived on open-circuit scuba, hadn't had a great time either. 'It's not the viz for me. It's the darkness. It was just so dark down there.'

I sat down opposite the third tech diver. He looked more cheerful than his mates. He reached into his mesh bag and produced a blue lobster.

'That's dinner sorted!' he said.

I smiled.

'Good dive?' he asked me.

'Ah . . . Not sure about that.'

He knew what I really meant: 'No.'

'Listen. Don't worry. Every day's a school day,' he said.

It was a nice way of looking at it.

The skipper emerged from the bridge. He must have overheard our conversation because he walked over and spoke to me.

'So you're telling me you had problems on your descent, and you and your buddy managed to sort them, that you were afraid, in the dark at eighty feet, and you then managed to do a controlled ascent with a completed free safety stop?'

He looked stern. I felt like a child being told off by his headmaster. 'Yes.' I almost said 'sir'.

He put his thumb up, stared seriously at me and nodded. 'Well done. Sounds to me like you did well.'

I needed that.

*

As we ploughed back to shore, I wondered if I'd salvaged anything of value from the wreck. Dad's warning came back: *Respect the sea, son. Cos it ain't going to respect you.* I had thought the meaning of those words had changed for me, but now I knew it was wise to hear them in the same tone forever, the tone he had spoken them in: serious and grave. Maybe I hadn't respected the sea enough, or maybe I had overestimated my right to explore it. Whatever my fault, the instruction was the same: take greater care. When I went to bed that night, Plato's Atlantis story came back to me again, this time with the chill of a personal admonition. As the sea let Atlantis rise and then brought it low, the sea has allowed your adventures, but it will take you down if you let it, and maybe even if you don't.

Perhaps it was enough that I had survived the *Fortuna*. The diver's maxim – that a successful dive is one from which every diver safely returns – was not just a platitude after all. It was hard-won wisdom, and it meant that I really could be content with the simple fact that I had returned. It was galling to think that such great risk might have been required to teach me a simple lesson, but I was grateful. *Come back alive. Your life is the greatest treasure you can bring back from the deep.*

There was solace in that thought. There was peace. It was Dad and Rán mixing again, my blent guide of the sea.

Before I turned over to go to sleep that night, I lifted my left wrist up to my mouth and licked the salt off the face of my watch. It tasted like a tiny victory.

The Three Nails

Cabo de Palos, Spain

In one legendary backstory of the Romany Gypsy people, our ancestor was a blacksmith who had forged the nails that were driven into the hands and feet of Jesus. There are several versions of the tale. In one, the blacksmith and his family were condemned to wander the nations because of their role in the torture of the son of God. In another, the smith was instructed to make four nails, but having been told in a dream that the fourth nail would be used to pierce the heart of Jesus, he handed over only three, and kept the fourth behind. The fourth nail became the talisman of the blacksmith's compassion for Christ. In this version, God gave the Gypsies the right to wander the Earth, and it was no curse; and he gave to them also the right to steal from the peoples they dwelt among, so that they would not be held to account for having broken the seventh commandment.

People have gone to their graves believing such things, as people have believed in the tale of Atlantis.

★

Cand and I had to return to Spain to carry out some administration of her late grandfather's estate. This would take me within thirty miles of the first beach where I had stepped into the sea and walked underwater. I couldn't go there without visiting Kees, the diver who had first invited me beneath the waves. Kees had probably forgotten about me, but I hadn't forgotten about him. Like the schoolteacher who with a single encouraging comment can spark the tinder of a dream, a career, a life, it was Kees whose matter-of-fact, *of course you can do it* attitude had first convinced me I would be able to dive without dying of fright, before Rán was able to come and ensnare me.

We spent several days cleaning the house and, because the utilities had been cut off to save on bills, ferrying gallon jugs of water from the nearest supermarket. With an old broom, many times repaired, I swept the red dust of the Sahara from the glossy white tiles of the terraces front and back. Everything was parched. Most of the cacti and succulents, left helpless in the open, had died. When the day came round that I was able to go diving, I had a deep thirst to be in the sea. Salt water it may be, but the sea can hydrate you in different ways. Being in it, at flight in the blue as a kestrel or skylark hangs in the sky, the mind is rehydrated, the soul is revivified: the memory of the sea floods back through the eyes and against the skin, and the dry earth of a scorched personality finds itself refreshed and ready for new growth.

Cand did not want to dive again – the claustrophobia of the mask and the difficulty of equalising played on her mind: memories as powerful for her as my recollections of the octopus, the nudibranchs, the beauty of the sea. But she rode with me over to Kees's shop in Orihuela to say hello.

In the taxi, flamenco music played at low volume. The female singer's voice trembled with the weight of generations. When we got to Kees's shop, he strolled out shirtless as before, removed his sunglasses, waved 'hi' to Cand and me, then, after a pause, reached out a hand towards me and firmly shook mine. A smile rose up from one corner of his mouth. He had remembered me after all.

'So you're really diving now,' he said. 'Good. You should be proud.'

I found that, with his permission, I suddenly was. Why not? The sun, the sea and Kees gave me their blessing and banished my self-flagellation. I felt the load of the last three years dissolve: the cold winter shallows off home, the months of lockdown without feeling so much as a lick of the sea, the mosaics of Baia swept clean of their broken shells, the brief nightmare of the steamship *Fortuna* and the relief of its conclusion. There was something very special about being back where I had started. I felt all the Atlantises I had chased converging, from Plato's vision of a many-channelled island to the complex world on the shell of a hermit crab. It was all so precious and none of these sea-things precluded any of the others. *I'm proud of you too, son. I'd rather you didn't keep doing this mad stuff, but I'm proud.* I no longer cared whether anyone thought there was rational cogency in the idea of dead people speaking to you inside. What was the difference between Dad 'really' speaking to me and me hearing things he said because so much of him had survived in me? I lacked any inclination to split that hair. If Atlantis could be real in a number of ways, then surely my dead dad could be too.

I arranged to come back the following morning for a dive at a location of Kees's choosing. The other divers and I turned up just before dawn and we all piled into Kees's van. As we neared Cartagena, red sunlight spilled across the Mar Menor, 'the Little Sea', and the hills and towers of La Manga were shadows against the blue beyond, an edgeless image of water and civilisation. The dive party exchanged names – most of us, anyway. These niceties don't always happen, but I prefer it when they do. A pair of buddies will always introduce themselves to each other, but it also feels good to say hello to everyone who is going below together. Once in the water, it's comforting to look around and see people you know by name.

There were two Irish sisters, Caroline and Sarah, who were used to exploring the clear but frigid waters of Donegal; an American man called Chris, who for many years had lived and worked in Spain, but had rarely dived there; and a heavily built Dutch motorcyclist with a beard and a Hells Angel T-shirt. On his right arm he had tattoos of skulls, a hawk moth and a pretty girl; his left was entirely covered by a blue octopus: on the one hand, terrestrial concerns, and on the other a life of the deep. I inferred something from this about the divided priorities of his brain hemispheres, and wondered if my need to keep entering the sea — sentimental, impractical, costly, and difficult to articulate — might also be emanating primarily from one side of my brain. It was born partly from a yearning to share moments with sea-creatures, and with what I increasingly felt to be the life of the sea as a whole; and partly from a quixotic need to encounter human things that had been lost in, and changed by, the water. Surely the pragmatic, precisionist, compartmentalising, food-seeking left hemisphere could not be behind all this. There was something powerfully of the ocean in the right side of the brain, something that had waited away the ages since our forebears emerged from the sea.

The van pulled over opposite the old lighthouse of Cabo de Palos — 'Lagoon Cape' — the end point of a little line of volcanic hills. The ground here was rougher than the terrain leading up to it: newer, less polished. People are born smooth and we grow gnarled with age, but the land is often born jagged and coarse and requires the kiss of time and the stroking of weather to calm its hackles. We got out and hauled our gear into the white light. The day had blossomed from chilly to very hot in the short time it had taken us to drive from Kees's shop to the dive site. As in Santorini, there was a heavy smell of sulphur in the air.

Pliny the Elder wrote that there was once a temple on the tip of the cape, sacred to the chief male god of Carthage, Baal Hammon, the spouse of the goddess Tanit. In later times the building may have been used for the worship of Saturn, the Roman god who mapped most readily on to the character of Baal Hammon: a weather-god who brought fertility and crops was easily replaced by one of agriculture, renewal, seasons and time. Humans reuse and rename gods as we redecorate and reinhabit the structures of the past. The sea brought the Romans ashore to the ruination of Carthage, but in small ways the destroyers could not help establishing continuities. 'Despite appearances, humanity is a coral reef,' the poet Don Paterson has said,[1] 'and we're mostly having the same thought.'

Three years had passed since I'd first 'kitted up' with Kees. I thought about that day, standing in a shimmer of heat haze as he buckled cylinders on to jackets and fiddled with spanners and Allen keys and pressure gauges and clips. As he strapped the equipment on to our daunted bodies I'd felt certain I would never learn to do this, and that I might not even make it into the water. I took a few breaths of cool air from the regulator. I was back, and I could do it by myself.

We followed Kees down a curling stairway cut into the slope that led to the beach, doing the trademark hunched walk of suited-up divers. In the past I'd have wanted to walk down the steps without using the wooden handrail, but I gripped it with my right hand and trod slowly and carefully. These days I was grateful for a bannister.

Similarly, I was glad to hold on to the shoulder of Chris the American as we paused on the beach before doing our final checks. Contented all was well, we filled our jackets with an excited hiss of air and walked out into the water. It was full of rounded rocks the size of little French loaves, and each wore a thin fur of algae, causing us to go gingerly. I was oddly fond of this typical hazard of Mediterranean dives. It ensures you take it easy on the way in, forces a slow pace. Once up to our chests we lay back and pulled on our fins. Caroline and Sarah were the first to roll over and sneak a glance below.

'Bloody hell, this is not like the sea off Donegal!' said Caroline.

'Told you,' said Sarah.

'I've already seen so many fish!'

'Yes. Welcome to the Mediterranean,' said Kees. I did feel welcomed. The Mediterranean gave a warmer embrace than the grey-green straits of home.

'Right. Let's go down and see what we can find. We stay together. OK?

As we went down the bubbles burst from the regulator in a spray of little glass marbles, and as they fidgeted their way upwards some of them coalesced into a much larger central bubble. This dominant bubble flattened as it rose, becoming a mushroom-like saucer of crystal. It had the appearance of one of the flying jellyfish from Riou and de Neuville's magical illustrations of *Twenty Thousand Leagues Under the Sea*. Expanding as they made for the surface, the big bubbles were reminiscent of the Portuguese man-o'-war, snub-headed, long strings of tiny bubbles trailing behind. When they hit the surface they made a big halo that hovered above every diver. Saints of the deep. Kees spun around to inspect the five of us and gave us a chipper two-handed OK. The viz was good, the sea warm and bright. *Where would you rather be?* I thought. Nowhere: this was the blue sublime, and as my eyes adjusted and began to interpret the bleary light I noticed that we weren't the only ones in the vicinity who were eagerly looking around.

A squadron of blue chromis fish hung beside us. They kept quite still, only occasionally flickering their fins to adjust their orientation. The chromis prefer to fine-tune their position, as opposed to hulking and muscling about the reef like the bulksome, menacing groupers. It feels pointless trying to describe the restless beauty of the colour of these fish. Where the charm of the various kinds of wrasse derives from the contrasting colours of their rainbow patterns, and their ability to surge forward with a ribbon-like grace and then suddenly halt with precision as though they had been swimming through a vacuum, the chromis's beauty is restrained and really a matter of the way its skin explores the possibilities of a limited range of colour. It exemplifies every hue in a single sector of the spectrum: a rainbow of blue. It is a deep sort of colour that drinks in light, giving it a density absent from everything else in the fish's surroundings, but it never stays the same for long, transitioning with inky smoothness from a rich purple to an intense lapis to a dark, matte navy. It feels like you are watching a colour wondering what it is: figuring itself out. The French author Sylvain Tesson wrote that 'if I were God, I would atomise myself into millions of facets so I could dwell in ice crystals, cedar needles, the sweat of women, the scales of spotted char, and the eyes of the lynx.' Were I God, I would live as the shifting blue of the Mediterranean chromis; and if God lives, then God must be there in the blue.

I found it hard to look away from the fish, as I must, and check my buddy was still nearby. I recognised I had been having highfalutin thoughts, which in the past had been a sign of gas narcosis. I checked my depth: we were too shallow for that to be an issue. It wasn't narcosis. Theological thoughts had stalked me, as they had Jonah, 'into the very heart of the seas'.

I had thought that fish did not have names, but if a name is simply the thing that others call you by, then the fish had names for everything: names of gesture and darting, of a short backward finning of suspicion, a turn of circumspection, or a sudden rocketing of pure fear. These are the names of the things they perceive, in our language octopus, grouper, or woman: names composed not of sound, but of motion. And in this language the names of things would overlap, so that when a diver causes panic they may share among the parrotfish, for a moment, the name of a conger or shark, or occasionally the name of a peaceful marine mammal – *essence of seal, but small*, perhaps. Normally though, as I believe I have noticed, divers have our own special name among many of the fish, and it is signalled by an intense eyeing and a readiness to bolt if disturbed, and if translated into human speech it would be something like, 'What on this watery earth are you?'

The names of other sea-creatures are easier for us to conceptualise, and to detect. We can prove that an animal makes a certain noise when it knows another specific animal is nearby. We can also prove that it recognises a sound that signifies itself: its own name.

We have known for some years now that naming happens in the sea, as it may have done for millions of years.[2] In 2013 the marine scientists Stephanie King and Vincent Janik of the University of St Andrews published their research into bottlenose dolphins and 'individual vocal labels' – in layman's terms, names. 'Bottlenose dolphins develop their own unique identity signal, the signature whistle,' King and Janik wrote.[3] 'Dolphins respond to hearing a copy of their own signature whistle by calling back. Animals did not respond to whistles that were not their own signature.' Put briefly: bottlenose dolphins appear to be the only mammals besides humans who use phonic names for each other – in their own language, of course.

It is possible that future research will uncover more examples, and if we include non-sonic modes of communication, then the matter becomes more open still. Much research has been done into the ability of primates to communicate using sign language, for instance, and sea-creatures might use non-vocal gestures to convey concepts as specific as individual names. And what of names relayed by means, and known to senses, that we may never be able to detect?

We were forty-five minutes in and nearing the shallows when I saw a pair of grey lines up ahead. A few kicks further on and there were more lines, forming what looked like a zany three-dimensional graph. Pockmarks of brown rust were visible on the grey, and ropes dangling in among the mess. It was a shipwreck. Kees hadn't mentioned we might see one, but as we got close I figured out why. This wasn't the sort of shipwreck people paid to go out and see. It was the remains of a small ship, more thin mast and rigging than hull, compressed and mangled by the water into what resembled a giant tangle of bent coat hangers. It didn't have much going for it. Unlike the big wrecks I'd seen elsewhere it didn't even seem to provide a haven for shy wildlife: there were a few chromis and bream swimming among the broken masts but no more than were hanging around in the open waters beside. This wreck was an eyesore: a snarl of cheap metal and line. It didn't make a pleasant sight in the moments before we resurfaced. On the beach I asked Kees about it.

'Yes, the catamaran. It sank about twenty years ago. It's not a pretty sight. But there's nothing to salvage, and here it's not a hazard to shipping. So I guess it will stay where it is. And we will watch it slowly . . . crumble away.'

Kees let out a short laugh that communicated everything but amusement. He was clearly upset. I remembered him three years ago in the bay of Cala Cortina, trying to gather up a brittle plastic bag as it crumbled away through the gaps in his fingers. Kees loved the sea, and he plied his trade in it, but he must also witness its despoiling. He looked glum as he shook the water out of his mask, like a guardian of the ocean who had no choice but to fight the long defeat.

★

When we got back to Kees's shop it was time to say goodbye for what might well be the last time. I said goodbye to Caroline, Sarah, Chris and the silent Dutch biker, then thanked Kees for starting me off on a journey that had taken me to places I'd only dreamt of.

'That's what diving has been to me,' I said. 'Not a sport. A portal into a dream. So thank you.'

He thought about it. 'I know what you mean. Good luck with the writing. Actually, wait here a second. I have something for you.'

Kees went into the shop and came back out a moment later. He held out a fist, fingers down. 'Here,' he said.

Into my upturned palm he dropped three pieces of metal, each of them rough, and coloured green and brown. They were copper nails, and the green patches were the scabs of oxidisation. The nails were curved into flowing shapes, punctuation marks of the sea, as if they had been listening to the water, as if by long submersion they had taken on its nature. They were beautiful, and it was because they were ruins: tiny products of human intention, painted unique by the artistry of time. They were of history, but they were redolent of Atlantis.

'They're probably not as old as they look,' said Kees. 'Sixteenth century maybe. But the small one could be older. The corrosion on the tip is different. See? It's so much more corroded. Could be a galvanised tip. Just possibly it's Roman. They sometimes made them like that.'

'I . . . Wow. Thank you. Where did you get them?' I asked.

'These are from the harbour of Cartagena. It's the only place around here where you will find anything like these.'

I asked Kees if he was sure he wanted to give me the nails. 'Of course. It's fine. I have others in the shop. I just thought you should have them.'

Kees couldn't have known what symbolism the nails contained for me. Even if he was aware of the story of the Gypsies forging the nails for the crucifixion, and I doubted it, I had not discussed my background with him at all. He had fished out from the Mediterranean a handful of totemic charms, three tiny things that had held ships together, and belonged to the seabed after those ships were gone. *See, he's only given you three*, I heard Dad say. *He's given you permission to steal Atlantis*. I laughed. So Dad was cracking jokes now, as well as issuing warnings. Perhaps these were also three nails in the coffin of my grief.

The nails also had a symbolism for me that lacked any connection to the Gypsies, Atlantis, or Dad. It was connected to the *Fortuna*. These nails had come out of shipwrecks that had long since faded away. They were stuck in them once, trapped aboard, prisoners of Rán. But they had persisted and come back to the surface, and now they were here in the sunlight, resting in my upturned hand. That was the real reason I'd cherish these objects. I too had escaped from a shipwreck. Over hundreds, perhaps even thousands of years, they had been on a journey that echoed the most frightening experience of my life. I could touch them now and remember how I had made it back to the surface. *Hard as nails*, said Dad. I laughed and shook my head as I stowed the three nails carefully away.

During the taxi ride back to the apartment, the driver tutted as he turned off the English-language news — the Queen, after an entire week, was still dead — to a station playing neo-flamenco. It was the folk music of this landscape, decorated with beeps and twangles, a good match for flying past the scorched vistas of old Spain on an elevated concrete road speckled with cat's eyes. It sounded like something very old overlaid with a veil of technology. It sounded like scuba diving, a modern riff on Atlantis.

The Ruins

Yonaguni, Japan

In 1986 a group of divers were exploring the ultramarine waters off the southernmost islands of Japan. This sea, frequented by hammerhead sharks, can be tranquil or typhoon-frothed, depending on the season. Off a southern cape of the little island of Yonaguni, the divers came across something that shocked them. Suddenly, their vision was filled by a remarkable rock formation that rose precipitously up from the ocean floor. A series of giant steps appeared to have been cut into it.

Could it be some kind of ancient ziggurat, built when this was dry land, during a previous ice age, perhaps? If so, it would catapult this marginal part of Japan to a status of global cultural prominence. This, after all, was the southern Ryukyu Islands, 1,200 miles from the electric spires of Tokyo. It was a quiet place of wandering ponies and modest populations. What if it had once been home to monumental builders comparable to those who had constructed the pyramids and temples of Egypt, Babylon and Central America? What if this was one of the most important and mysterious ancient sites in the entire world?

There are certain edifices that we find it hard to believe occurred by chance. We fall prey to a version of the intelligent-design argument; but we believe humans, not God, were the geniuses responsible.

The Yonaguni structures tell us that at times we can't help seeing a human hand where there isn't one: but why?

With their steeply sloping sides and broad, flat tops, the Yonaguni rock formations resemble the buildings in the painting *The Last of Atlantis* by the Russian artist Nicholas Roerich. Roerich's artistic vision was clearly inspired by his time in the Himalaya. There he painted Tibetan vernacular architecture, including buildings with sloping trapezoidal walls that appeared to have risen in relief from the very ground beneath them. This was human design as an aesthetic extension of the rock it sat upon. And when Roerich looked at it, he didn't just see Tibet. He also saw Atlantis.

Some who have dived at Yonaguni and spent a long time examining its rock formations have perceived in them what they regard as undeniable hallmarks of human intervention. They include the geologist Masaaki Kimura and the writer Graham Hancock. Hancock is writing partly in the tradition of the Swiss author Erich von Däniken, who challenged mainstream narratives of ancient human culture by positing early transcontinental contacts and the influence of divine forces or, for von Däniken, aliens.[1] For Hancock, the Yonaguni structures fit into a theory of how complex ancient civilisations were flooded during the melting events that accompanied the end of the last ice age. He recalls how, on one dive:

> *Despite lingering doubts, I felt a sudden surge of confidence that nature could not have done it – not unaided anyway. On the contrary, the pattern was a complex and a purposive one, rather difficult to execute in any kind of rock, and the more I studied it the more obvious it seemed that it was deliberate and planned.*[2]

The assertion that an underwater formation was deliberately human-made must be based on more than an intuition. Some underwater rock formations that look 'deliberate and planned' have turned out to be exactly that, but convincing proof is required. In 2024, researchers from Germany's Kiel University identified an assemblage of large stones on the seafloor under seventy feet of seawater, about six miles off the Baltic coast of Rostock District. The stones got the scientists' attention because they were arranged in a line more than half a mile long and stood out in an area otherwise largely clear of similar objects.

Some were boulders too large to have been moved by ancient people, but the majority certainly could have been. The researchers concluded that this was the remains of a Stone Age wall, mostly around three feet high, constructed by Mesolithic humans using the largest boulders as their initial plot points, which were then connected using the smaller stones. The wall would likely have been used by hunters to funnel wild game into a trap, such as boggy ground or shallow water, where they could be more easily dispatched. The Baltic wall resembles the 'desert kite' structures of south-west Asia, which are similar in appearance, and are thought to have been used for trapping wild gazelles.

Such proofs and rationale are lacking in the case of Yonaguni, which is why most geologists disagree with Hancock's assessment of it. They think the 'monument' is simply a naturally occurring structure with some qualities that might appear at first glance to be man-made. Robert Schoch, a professor of natural sciences at Boston University, seen by some as an 'alternative archaeologist',[3] has pointed out that the formations are composed of separated bedrock, rather than being built from separate blocks. They are made from sedimentary rock, the horizontal cleavage within which means it breaks along parallel lines. Another key to interpreting Yonaguni is to remember that it sits on the edge of the Philippine tectonic plate, a region of regular seismic disturbance. This causes rock to split vertically. Combine these factors and add the smoothing effects of marine erosion over time – and the illusion is complete.

The Yonaguni rock formations are extremely photogenic; arguably more visually compelling as 'structures' than the 'hunter's wall' of the Baltic. Pictures of the 'monument' taken from certain angles can create the impression of something like a Neolithic assemblage of megaliths; perhaps even a lookalike Mayan temple submerged in deep-blue water. At points the similarity is uncanny. Unaccompanied by deeper enquiry, it is tempting to presume that it must have been built by humans. Mix this with the fact that the academic establishment can be elitist , and at times has vested interests in maintaining narratives that suppress the truth, and you have all the ingredients needed to attract controversy.

The problem with inferring human design without properly

assembled evidence is that resemblance of form does not guarantee a common origin. On planet Earth, things look like other things. Some types of tabulate corals look uncannily like little model blocks of flats, and blocks of flats are themselves analogous to coral: they are rock structures built by soft organisms to protect and house themselves. Sodium chloride – common salt – forms natural crystals in the shape of pyramids. We know which ones came first, but it is hard to look at a pyramidal crystal and not be put in mind of the giant buildings of ancient Egypt. What does all this tell us, if not that we are simply part of an organic feedback loop? We are inspired by the shapes around us and ancient human structures were often designed to imitate nature; therein lies the inherent circularity of any argument of intelligent design. No wonder the huge outcrop of stone shapes that protrudes from the seabed off the Japanese Ryukyu Islands is known as the Yonaguni Monument. That's what it looks like: a monument.

Seeing things underwater, the muses of the sea can seize our minds. We get high on the blue, and summon inspirations. In his tale of Atlantis, Plato invokes Mnemosyne, the Greek goddess of memory, to aid in the telling of the story. He's giving his audience a clue that something mysterious might be afoot. A straightforward account would require no muse.

★

At the age of twenty-one I set off from the coast and embarked on a disastrous year-long expedition to find my worth in the city. The strength of my degree, hard-earned and prestigious, had convinced me that I would only have to land there and the victory would be mine. There had been a promising start. I made inroads, went for high ground. Two companies vied for my services, inviting me forward with offers of long hours performing difficult mental tasks for decent money. I became stressed. Pallid, wired and unable to sleep, I cast my nets over the urban night and pulled in the flotsam of bad habits. Far from the sea, cut off from the vital supply lines of family and rest, I was caught in a pincer movement, hammered between unrealistic expectations for myself and the Roman anvil of London.

Thinned and depressed, my self-image in ruins, I returned home, washed like a river back to the sea. In a matter of days I went from courting my golden future to shivering on my parents' couch in the foetal shape of a prawn. It was the dog-days of summer but I felt freeze-dried. The simplest activities, such as making a cup of tea or pulling on socks, were suddenly daunting and indescribably complex. The joints and pistons of my conviction had seized up. I could not flow around or into things but only sit – or more often lie down – near them, and wait.

Some animals are able to survive long periods of drought by aestivating. Aestivation is a period of dormancy, akin to hibernation. The systems of the body are slowed, necessary chemical sacrifices made. But while hibernation is usually a means of surviving through winter, lowering body temperature, breathing speed and metabolic rate to eke out energy until the sun returns, aestivation takes place in the summer, and is a method for coping with dry times. *Austrothelphusa transversa*, the Australian inland crab, is found across wide areas of Australia including arid central regions. It survives dry spells by burrowing into the soils close to the water table, where the sediment or clay contains traces of moisture, and it can lie still in a darkened chamber where conditions are relatively benign for a crab. Once installed there, it aestivates, having plugged the entrance to its burrow so that the dampness in the air cannot escape. It waits. It will remain installed in its pocket of earth, its womb of clay, until the rains fall and replenish the disappeared rivers and pools that only exist in the brief wet season. They are known as ephemeral rivers. 'Ephemeral' comes from the Greek word ἐφήμερος – *living but for a day*. The crab's only hope of rescue is in the return of the waters.

My mum thought regular visits to the sea would do me good. 'Come on, son. We are going down the seafront,' she said, as if it were a military front. I abandoned myself to the waves of persuasion. It worked. Whereas I always woke up feeling the same way, blank and depressed, I observed that the sea was unlike me. Resembling healthier people, it had moods. An exhilarating range of them:

chilled; catatonic; rippling with ecstasy; frothing with rage; crazy wind-whipped spiralling party mode.

It smashed itself to pieces one day and sat still the next. On Saturday it might turn itself into a varnished portrait of the sky, and the next day sculpt itself into a turmoil of froth-capped bulges, brown, the blue gloss gone. It was always the sea, though. It could be and do all these things and still be itself.

The ocean was teaching me something. You can journey into your own future without maintaining a resemblance to your self of yesterday. In fact, abandoning that self can work as a kind of propulsion. The nautilus, descendent of a staggeringly ancient lineage of beings, moves by abandoning water behind it, by strongly and repeatedly letting go. Blind to what lies in its direction of travel, the nautilus relinquishes its way forward. It embodies an ancient Greek view of the future as that which awaits us behind our backs, into which we must move unawares. Deep oceanic time has tested this way of life and found it workable for 500 million years. I saw that I must release not only the past, but the notion of my life as a planned structure. Many forces play vital roles in our making: hard currents rasp at us, new shapes emerge, and an edifice is revealed, often largely independent of any prescient design. The sea of circumstance is our maker, in which our intention is but one power in the mix.

Whatever its origin, the Yonaguni 'monument' is at the very least a monument to itself. Monumental in scale, built by the forces of stone and the sea, it is a statement of the nature of Earth, designed by rock and the ocean. Whether or not we enjoy talk of spirits and gods, we owe it to the world to roll back our notion of creativity to a stage prior to ourselves. If a human structure is remarkable because it evidences creativity, then any structure, built by that same principle working through something else, must surely be as marvellous.

My fetish for human ruins had been expanded. Everything had become a ruin now. By jettisoning the obsession with the human, a new vision of the seabed, the precious seabed, can be born. Every inch of it has potential, made of the old and about to play host to the new. Remaking its own myth, an old story out of which infinite things can be read.

11:43 a.m., 7 June 1692

Port Royal, Jamaica

We stepped through the terminal doors of Norman Manley Airport into a wall of Caribbean heat. On the taxi rank people fanned themselves with boarding passes as cars rolled along in first gear. A few men in shirts, slacks and smart shoes ambled around selling cab rides to sun-struck arrivals. Across the road, a trader at a small stall made from sticks and tarpaulin was slicing the tops off green coconuts, jabbing straws into them, and selling them to travellers weary and thirsty from the plane.

An erratic man rushed up to us offering a ride. He had coiled silver hair and eyes that looked through my head to some wild place that lay far beyond me. I declined.

A few moments later a second man walked calmly up. 'Hello. Taxi?' He had a slow and kindly manner that suggested that he was no keener to take us than someone else. We got in the car and he introduced himself as Dwight. 'First trip to Jamaica?' he asked.

'Yep,' I said.

'Welcome! You will have a nice time.' It sounded like both prophecy and command. 'I saw that you brought some mosquito repellant. Good work. Because wait until six. They will eat you alive!'

'Palisadoes' is the old Portuguese name for the natural but unlikely-looking harbour arm of Kingston, a slender tentacle of land that dangles westward into the Caribbean Sea, cordoning off one of the largest natural shipping havens on Earth. Dwight drove us along a road flanked by ten-foot cacti as a warm breeze swept through the windows of the car. To the north of the road are little peninsulas of mangrove swamp, and at the far end sits Port Royal. Founded by Spanish colonists in 1494, enjoying a strategic location at a nexus of key sailing routes, Port Royal grew swiftly in importance. By the seventeenth century it had become, though tiny by global standards, the commercial centre of the entire Caribbean, acquiring a second reputation as a hothouse of debauchery. Privateers based there were given incentives to harass and steal from Spanish ships, spending their earnings on vice. What Port Royal lacked in size it made up for in alcohol and sex workers. Its nicknames included *the wickedest city on Earth* and *the Sodom of the New World*. So when on 7 June 1692 an earthquake and tsunami destroyed a large part of the town, there were bound to be those who suspected God's hand was involved. A third of the population died in a couple of hours. Destabilised by the tremors, water welled up from the ground as though conjuring itself out of nowhere, and buildings, carts and people began to sink into a kind of quicksand. The liquefied area contained the city cemetery, from which corpses in various states of decay sprang up to the surface. They bobbed about alongside the living who screamed and swam for their lives.

We got out of Dwight's taxi at the old Morgan's Harbour Hotel. In a bush by the front doors a blue-green hummingbird thrummed as it inspected long orange flowers. A hot wind blew through the white stone lobby, bringing with it the smell of Kingston Harbour – a humid peppery musk of salt water with a faint tang of diesel.

The door to our room was guarded by tiny lizards. They clung to the wall and when I turned the key in the door they disappeared in a flash to the other end of the gangway. The room had a small balcony from which, in the distance, the Blue Mountains could be seen, ringed with smoke-like cloud. In front of them lay Kingston Harbour, wide and calm, and above its flat and sultry water the pelicans patrolled in pairs and threes. They glided lazily before spotting a fish, at which they would ramp upwards and, in a feat of self-origami, fold into shears and smash down into the water. They were hunting in the area where Fort Carlisle had been before it sank. Fish swam in the place of soldiers, sailors and the enslaved.

The Port Royal disaster is rare among historical events of pre-electric times in having been dated not just to the day, but possibly to the minute. For this we can thank two key people: an anonymous Port Royalian who once owned a pocket watch, and the scuba diver Edwin Link. Throughout the 1960s, Link made a series of expeditions to the submerged ruins of the town. He returned with many artefacts, most famously a wadge of sea-blued metal that turned out to be a corroded brass pocket watch. In spite of its terribly damaged state, analysts managed to use X-rays to determine that it had been made by a French resident of the Netherlands, Paul Blondel, and that the hands had stopped moving at 11:43. Presuming it was in the pocket of somebody who succumbed to the earthquake, the watch must have entered the water mere seconds before this time, as it would have been unprotected from moisture by such later inventions as greased gaskets or a screw-down winding crown. It was Link's most famous find, and one of the only times in history that a diver has been pleased to discover a watch wasn't waterproof.

*

Across the street from the hotel there was a young man with a silvered eye selling juice drinks from a small cart. He sat on a rocking chair as, beside him, a young woman plaited crimson strands into her hair. I bought a grape soda and sipped it: it contained a cupful of sugar. In the damp heat its intense sweetness made more sense than it would have in the cold of home. I felt instantly refreshed.

I asked the young man if he was from Port Royal.

'Yah, man!' he answered. 'Henceforth you come and buy your drinks from me.'

His archaic English speech seemed a relic of a departed age. He reached out a fist and I lightly touched it with mine.

I had sent enquiries to a couple of local divers but after initially responding to my messages they had gone quiet. We decided to go and see the surviving parts of old Port Royal. The place was famous for being drowned, but of equal interest to me was the fact that not all of it had been. We booked a tour around Fort Charles and were guided by Dane, a small man with kind eyes, cornrows and an encyclopaedic knowledge of Jamaica's colonial period. As I listened to him talk, I thought about my childhood in England playing with toy pirates and redcoats, thinking it was the coolest thing in the world. They left out all the enslaved people and plantation workers, all the instruments of torture and capital punishment. There were cannons and swords but they suggested the possibility of a fair fight. My toy had been as much a model utopia as a model of Port Royal. Had I been old enough at the time for someone to tell me that the pirates were mostly men who had fled the brutal regime of naval service, or that forts like this were built on the trade in human beings, shipped under decks as though they were cargo?

'There is a lot of hidden beauty in Port Royal,' said Dane, before explaining a history that in many respects was far from beautiful. Fort Charles was, in its way, a beautiful building, but only because its brutality – the effectiveness of its military design – was superannuated now. It was clothed not just in an overgrowth of plants but with the cunning disguise of time. As guardian fortress to the mouth of Kingston Harbour, it was for several hundred years one of the most important entrenched positions in the British Empire. As Dane showed us around he told us about Britain's conflict with the Jamaican Maroons, whose name probably derived from the Spanish *cimarrón* meaning 'wild' or 'untamed'. The Maroons were the descendants of African Jamaicans who had liberated themselves[1] from slavery, and who likely absorbed influences from the remaining indigenous Taíno people of the island. In the eighteenth century, they had fought guerrilla wars against the British, the first lasting twelve years. Led by Nanny of the Maroons (whose face appears on the Jamaican $500 bill) and her male co-commanders Quao, Accompong and Cudjoe, they were heavily outnumbered and outmatched in terms of weaponry. In spite of this they defeated the British through a skilled, prolonged campaign. The largest maritime empire in world history was outdone by a tiny, free force who fought only on land and for the sake of their own liberation.

The resonance was obvious. Dane was telling me the story of Atlantis and primeval Athens. The difference was that this version was true. In his telling, Britain was the power-hungry Atlantean Empire, and the heroic Athenians were the Jamaican Maroons. It was an exhilarating thought, tendering a non-Eurocentric reading of the *Timaeus* and *Critias*, texts that had been used to justify the colonial 'project'. Like all analogies made to the Atlantis story, it wasn't a perfect fit. Still, perhaps Plato's writing could play a vital role in the modern world: hold a mirror up to history and politics, and examine where the seats of evil lie. For us to use it for its original purpose, though, the parable would have to regain primacy over the aesthetic kaleidoscope 'Atlantis' had come to mean, and that was unlikely. Could its meaning ever transform from 'kingdom under the sea' to 'rich, ill-fated maritime hegemon' in the modern mind? Probably not. But standing in Fort Charles, listening to Dane, it felt more relevant than ever.

'Come with me,' Dane beckoned. He led us over to a large walnut tree and knelt down in its shade. He searched around on the floor and gathered a couple of walnuts, then smashed them on the flat of a large brick. He doled out the contents and we ate.

'A little Jamaican hospitality. From me and the walnut tree,' said Dane.

*

On the boardwalk three ships were at anchor: the *Abraham*, the *Loose Cannon* and the *Persistence*. In the water between the boats, its surface rainbowed here and there by diesel, groups of skinny trumpetfish hovered in threes and fours, slim silver swords of light. A bar served rum and steamed fish below the flags of Jamaica, Canada, Australia, and the Jolly Roger.

'Damian?' I heard someone say behind me. It was Llewelyn Meggs, a local divemaster and marine biologist I'd arranged to meet.

Handsome and kind-eyed, he had a warm smile and a swimmer's physique. Llewelyn had helped map the remains of the sunken half of Port Royal. If anyone could persuade the authorities to let me dive in the pirate city, it was him.

Of course, it was not that simple. The sunken half of Port Royal is unique – the only drowned city of its kind, not just in the Caribbean, but in the entire Western Hemisphere. The Jamaica National Heritage Trust has applied for it to be listed as a UNESCO World Heritage Site. Ever since it hit the seabed, Port Royal has never been left alone for long, and because its remains are so close to the surface, access has never been a problem for those who saw a chance to enrich themselves at its expense. In the immediate aftermath of the disaster, looters and opportunists scoured the freshly submerged city for gold and other treasure, and in the years that followed, professional treasure hunters known as 'wrackers' – some of them based over a thousand miles away on Bermuda – ransacked the remains. Attempts at recovering valuable items, some announced and others done in secret, continued sporadically until the twentieth century. The American diver Robert Marx led a series of organised expeditions to salvage the artefacts that had survived nearly 300 years of incursions from above, either because they had been ignored as worthless, or because they were well hidden or had otherwise been missed. Marx himself has a mixed reputation, regarded by some as a pioneer of marine archaeology, and by others as a somewhat ruthless treasure hunter. Whether or not there is some validity in both of these views, what is undoubtedly true is that Robert Marx was not Jamaican, and that as such his removal of thousands of significant items from the site has left a bitter aftertaste in the mouth of the country's cultural conservators. I pondered the fact that while in England I was part of the Romany Gypsy community, a socially and politically maligned ethnic group, here I was a white Englishman, destined to be unaware of historical and emotional sensitivities over which he might be trampling. Here I was bound to my ignorance, and it was more than a lack of knowledge: it was a presence. It pulsed like a fever. I thought of the words of Jamaica Kincaid in *A Small Place*: 'The thing you always suspected about yourself the minute you became a tourist is true: a tourist is an ugly human being.'[2]

For this mixture of reasons I was unsurprised when Llewelyn explained that he couldn't offer any assurances about whether he could help me secure permission to dive in the ruins.

'It's gonna be touch and go,' he said. 'All we can do is explain what you're doing, be humble and hope for the best.'

In the meantime there were a couple of dives we could do outside the harbour, in the Port Royal Cays. Llewelyn suggested Farewell Buoy, which marks the point where ships depart the shallows off Kingston; and the wreck of the *Cayman Trader*, the heavily broken-up remnants of a ship that caught fire and sank in 1977.

The next morning we motored out into the Cays with the sun on our backs and the cool sea foam of the harbour flung behind.

'All right,' said Llewelyn. 'The maximum depth on this dive is a hundred feet. So we're going to drop to the slope and explore the environment. We'll descend with the chain as a reference, but keep your distance from the chain. It's covered in fire coral. Not only is it precious, and takes a long time to grow, but it will give you a hell of a rash if you touch it.'

On the gentle slope of the seabed there were huge sea fans. I stared at them in amazement, these intricate skeletal doilies of life which appear as mere vertical sticks of coral when viewed from the side. I looked at Llewelyn. He nodded as if to say, *Welcome to the Caribbean Sea*. Small reef fish – black and yellow reef butterflies, juvenile angel-fish and tang – flitted among the sea fans and stunted corals. It was a beautiful, tranquil scene, but I was unprepared for the living work of art that was about to swim into it. Llewelyn waved. He had spotted something coming up out of the purple half-light of the depths. It was a queen triggerfish, also known as an 'old wife'. As its fins moved gently, undulating with telltale ribboning, it rose slowly. Llewelyn and I hung stunned in the water at the luxurious cobalt blue of its body, broken up by the yellow of its hindquarter stripes and the pale patch of its belly. It did not hang around for long, soon melting off into the shadows.

We turned to explore a seascape of fat barrel sponges and brain coral, where spotted drum – *Equetus punctatus*, flamboyant little zebras of the reef – picked away at the rocks. A large concrete block was smothered in flamingo tongue sea snails, tiny fashionistas covered in asymmetrical yellow spots delicately ringed in black.

As we drifted around I relished the improvement in my ability to move up and down by controlling the amount of air in my lungs, rarely touching my dump valves or the inflate and deflate buttons on my BCD. The trouble with a buoyancy control device is that, unlike a swim bladder or the lungs of a marine mammal, it sits outside the body. You operate it by hand, pressing a button to let air in and out, unable to sense, within yourself, what it is doing. The result is that using it, while effective, is a staggered process. The diver perceives that something needs altering, uses a gadget, then waits to see whether they got it right and, if necessary, repeats the steps again. Over time this process starts to become second nature, but it can never replicate the biological intuition of using your own organs to achieve the same effect, no matter how ergonomic and well engineered the equipment. As I became more experienced, I'd learnt to use breath control. Moving up and down by carefully changing the amount of air in your lungs is in many circumstances preferable to adjusting the gas in your drysuit or BCD: it's immediate, intuitive, precise. It is a little closer to moving through water the way a sea-creature does.

Llewelyn signalled that I should stay put while he went up to check the location of the boat. I hung still in the water, and was suddenly hit by astonishment: that I, who had always been terrified of the sea, of drowning, of the endless blue, was currently suspended at a depth of a hundred feet on the other side of the world. I was not only calm, but utterly content, until I started to wonder whether Llewelyn was coming back. Suddenly I felt a tap on the shoulder and turned around to see Llewelyn waving. It was time for us to ascend.

After a short surface interval on Lime Cay, a fleck of white sand held together by mangrove trees and the tide, we were to dive the *Cayman Trader*. We dropped anchor beside the wreck and Llewelyn went down to secure it. He was fifty feet below, but I could see him as clearly as though through hazy air. When he popped back up he hollered two words: 'Nurse shark!' Much of the meaning in the word 'shark' hangs on the intonation, but Llewelyn sounded joyful.

Phew. I wouldn't become aware until later that nurse sharks rank as high as fourth in the 'bite frequency' list of shark species that have injured humans. This, though, is put down to unwise attempts by divers to approach or pet nurse sharks which, although peaceful animals, do not always appreciate an unsolicited advance.

I descended behind Llewelyn and we started to swim around the wreck, which was severely broken up and scattered over a wide area like a fallen, burnt-out building. He indicated that in the distance, maybe seventy feet away, a trio of eagle rays were circling slowly like ghost birds. As their fins rippled easily through the water, they had the look of cloths dropped through still air. Llewelyn made a V with two fingers, brought them up to his eyes, then pointed them towards the rays. *Take it all in. Not everyone gets to see this.*

There were fish everywhere: jacks, porcupine fish, yellow reef fish of various kinds. We swam through the wreck and entered a metal box which had been a square room below decks. In it were perhaps a thousand small fish, and the room was tilted such that one of its top corners was now its uppermost part. There was a silvery pocket of air trapped there, and the bubbles that streamed from our regulators were sliding up the sides of the roof to merge with it. Llewelyn signalled *watch me*. He drifted calmly into the apex of the room, exhaled a bloom of bubbles, removed his regulator, and breathed in a deep lungful of air from the bubble trapped in the peak. He turned back towards me and grinned, then blew out the air. It looked like he had taken in a draught of mercury and released it into the sea.

Llewelyn replaced his regulator and swam off. I decided that since I would probably never find myself here, or perhaps in any sort of shipwreck, ever again, I must attempt to copy what he had done. I swam up to the bubble, took out my reg, and managed to sup a small mouthful of air before realising I was terrified. I jammed my regulator back in my mouth and swam away after Llewelyn.

On the other side of the wreck a grey shape was moving over the sand: a ghost on patrol. It was the nurse shark Llewelyn had spotted earlier, slowly undulating its way over the rusted remains of a gunwale. Nurse sharks have been around for 100 million years, and this one moved so relaxedly it looked like it might live forever if it could only be left alone. With its easy movement and wing-like pectoral fins, the shark swooped softly over the wreck with a hypnotic, sinuous snaking. Nurse sharks can reach ten feet long but this one was smaller. Their teeth – little cream-coloured thorns, angled backwards to stop prey escaping – cannot be seen from a distance. I could just make out its fleshy barbells, the white chemosensory appendages beside its mouth with their look of a vampire's fangs. The tiny pale eyes peered from their wide positions on its broad, flat, almost catfish-like head. The shark's swimming betrayed not a hint of unnecessary motion. It moved as if water within the water, and after a moment its grey resolved back into the milky limits of our vision and it was gone.

*

That evening, over a short coffee at the hotel bar, I spoke to Lamoya, a senior member of staff. She patrolled the hotel with a measured gait, shoulders back and braids neatly draped. I perceived that the 'all is well' vibe of the place was in large part down to her.

As we chatted I brought out the sketchy map of Port Royal I'd drawn in my notebook, which showed the present coastline with a dotted line representing where the shore had been before the 1692 earthquake.

'You drew this?' asked Lamoya. 'It's beautiful, you know.' She traced a finger along the little streets of ink, and off into the schematic waves of the sea.

'I'm pretty excited to be here. I've wanted to come since I was small.'

'Really? Specifically here?'

'Yeah . . . I had a model of Port Royal when I was a kid.'

'What? A model Port Royal? And now you're here, what are you gonna do?'

'The main thing I want to do is dive in the sunken city. But it might not happen.'

'What do you mean it might not happen?'

'Well, it's a sensitive site. It's been exploited over the years . . .

Right after the earthquake it was ransacked. And some of the excavations since have been controversial. They're cautious about who gets to dive here. Which is totally fair enough.'

'But you flew from England.'

'I did, yeah.'

'And they're saying you might not be able to dive?'

'I did get out diving today. We saw some amazing things.'

'But the *main* thing you came here to see?'

'Yeah, it's tricky . . . It's fine if I can't do it, though. I'll understand.'

Lamoya threw back her head and laughed a mighty, unselfconscious laugh. 'Listen, whoever you're dealing with over this, you need to give him the full hundred.'

'The full hundred?'

'Yah, man! Damian. Let me tell you something about Jamaican people. They aren't going to just help you automatically, if they can't perceive the reason. You come here just casually, like "Yes, I maybe want to go diving", or some such ting. That kind of vagueness. What is that! That's not going to get you very far, my friend.'

'OK. So I . . . give them the full hundred?' I still had no idea what it meant.

'Ha! Yes. What you need to do, is to tell the person who can help you, your diver man or whoever, what you just explained to me. That you dreamt about coming to Port Royal since you were a child, that you have a passion to see the underwater city. That you crossed over these thousands of miles of ocean specifically to see what is in that water over there. And if you go home without having done this thing, it will be one of the disappointments of your life. You came to Jamaica and missed your dream? Na, man! It cannot be possible. This kind of explanation, I guarantee you this man or woman you talking to will fully understand. And they will help you then. But you have to tell them, Damian. They will not know automatically. You got to talk to them like this. You got to give them the full hundred.'

Lamoya handed me back my notebook. 'Good luck,' she said. 'Now go and talk to your man.' She swung on her heel and walked out into the sunlight and her business.

Bearing Lamoya's advice in mind, I stayed up late and reworked my application to dive in the sunken city, cramming the maximum permitted number of words into the form which Llewelyn was going to send to the National Heritage Trust. I also explained to Llewelyn, in similar terms to those Lamoya had used, how much it would mean to me to dive the site. Across the steel of his seasoned divemaster's mien I saw a flash of empathy for the simplicity of my desire.

*

Forty-six hours before I was due to leave Jamaica, I got a voice message through from Llewelyn. The Jamaica National Heritage Trust had approved my application to dive in the sunken city. I was stunned. In spite of having taken Lamoya's advice and made as decent a case as I could for myself, I'd made my peace with the fact that I might well leave without having seen the underwater ruins. It was a privilege to have come here anyway. I had seen Fort Charles; met people who – in my fancy, but surely in some cases also in reality – were descendants of the earthquake survivors; stared at the waters above the ruins; and dived with a man who had mapped them. This was more than enough. But something had now happened that was better than more than enough. I raced to withdraw some American dollars to make a donation to JNHT's heritage work, a fair request in partial exchange for their permission.

The next day Llewelyn and I met early. The pale-gold morning sunshine flooded Kingston Harbour with light. By the shoreline we met Mr Selvenious Walters, deputy technical director of archaeology at the Jamaica National Heritage Trust. I noted his folded arms and stern expression. He didn't speak until Llewelyn addressed him, as 'Mr Walters'. I began to voice my effusive thanks to JNHT for authorising the dive.

Mr Walters stared into my eyes and waited for me to finish. 'You said in your application that this dive would be for visual observation only,' he said. 'I see you have brought your camera.'

'Ah, my apologies, Mr Walters – I did presume that would mean I could take photographs for my personal reference. By "visual observation" I was emphasising that I will keep my distance and touch nothing onsite. I won't be filming any video.'

'Hmm,' said Mr Walters.

'I apologise, Mr Walters. Of course I'll be happy to leave my camera onshore.'

There was a brief silence.

'I mean, you reckon it's OK if he takes a few snaps?' asked Llewelyn.

Mr Walters remained silent, his face unchanged.

'Under normal circumstances, had we known you intended to take any pictures, we would request an additional fee and a more thorough description of your work plan.'

'I understand.'

'But since you are diving under Llewelyn's supervision, and you are telling me any photographs you take will be solely for personal use, I am happy to grant you permission,' he said.

'Ah, thank you so much, Mr Walters. That's incredibly kind.'

'Enjoy your dive,' he replied. 'I will wait here for you onshore.'

Llewelyn turned to me and gave me a wink. We got our gear out of his truck and kitted up. It was actually happening now.

Our entry point would be the remains of the wharf by the Old Naval Hospital. This earthquake-proof structure was prefabricated in England in the nineteenth century before being shipped to Jamaica in pieces where it was erected and concreted into the ground. It is still the strongest building in the neighbourhood, and the muster point for Port Royal's inhabitants in the event of a tremor. An image of Robert Marx crept back into my mind: I had seen him setting off for the ruins from this spot in a crackly 1980s documentary on the internet. A double-edged shard of feeling went through me, piercing the moment: my elation on one side, discomfort on the other. I told myself that if JNHT hadn't thought my request reasonable, then they would have told me so. *I can't be that bad a bloke*, I thought. Dad used to say that when trying to reassure himself. I couldn't tell if I heard it in my voice or his.

We entered the shallows where the bricks of the old wharf sat just below the surface, taking care to step in between the poisonous black sea urchins.

'Is the sting bad?' I asked Llewelyn.

'Yeah. That's really gonna ruin your day! Ha ha,' he said.

At a speed that looks like stillness to human eyes, the urchins made their rounds over the grey flagstones of the wharf. They were like living sea-mines, an ancient and inscrutable form of life sharing traits with the hedgehog, starfish, seedpod, snail. Their body exhibits the fivefold symmetry of all echinoderms, in whom nature has sculpted a tribute to the perfection of odd prime numbers. Like starfish, the first sea urchins appeared in the Ordovician period, around 450 million years ago. To put this age into perspective, the earliest dinosaur fossils date to the Triassic period, between 228 and 245 million years ago. Urchins have been around almost seven times the span of years that separates us from the tyrannosaurus, so my desire to step around them was born out of both self-preservation and a wary respect.

'All right. What we're gonna do is swim out to the marker you can see there. Then we're gonna descend. We're looking at a maximum depth of maybe twenty, twenty-five feet.'

I took a look below. The viz was poor, a bitty soup of grey and blue.

'Not looking too great, I'm afraid,' said Llewelyn. 'But we'll see what we can see.'

I didn't care too much: the thrill was coursing through me, and the mist in the water added mystique to the dive. The fact I couldn't see the ruins from the surface had ratcheted up both nerves and excitement. We were hovering above the sunken city now, but only on the seabed would there be a revelation.

As we descended down the marker pole the fog cleared somewhat. The first thing I could make out on the bottom were the fat shapes of sea cucumbers, dark brown against the khaki sand. I looked to my side and gave an OK to Llewelyn. Slowly, he pointed out something behind me, the way a person might indicate a bear. I turned in the direction of Llewelyn's outstretched hand. And there it was: the corner of Fort James.

The brick structure in front of me had been sitting below the water for 330 years. This L-shape, mortared together by men in many cases stolen from their families and bound for a hook of land 5,000 miles from home, had endured. Facing up to it, I realised that it wasn't simply a sunken city. It was a mausoleum. What was this span of time? It was not so very long ago, and it was long ago, all the same. It would be a whole century after the sinking of Port Royal before Europeans built a permanent settlement in Australia, and another still before human feet pressed on to the cold rock and ice of Antarctica. The pirate city went below in the age of sail and horseback, seventeen decades before the internal combustion engine was invented and when anthropogenic climate change would have seemed an absurd idea. Yet only a dozen generations had passed, if that. Twelve steps from the pirates to today's Port Royalians. That didn't seem like many at all.

The fort's corner stood proud of the seabed: strong, still hard to assail. I handed my camera to Llewelyn and motioned with the other hand, requesting a photo. He took the camera and, remaining perfectly stationary in the water like a fakir of the sea, got a couple of shots. The lines of a smile were visible around his regulator. Llewelyn could tell I was rapt. He looked pleased. I knew pictures couldn't do the experience justice but I wanted a record of this moment, the end of a winding road from a toy set thirty years ago, to the actual walls of sunken Fort James. I was swimming through a memory: Jamaica's and the sea's, but also, in its small way, my own.

Halisarca caerulea, the thin encrusting sponge, is able to grow on glass. It coated the many bottles that were strewn about the sunken city. Flesh-coloured and scored with little star shapes, it made the things it covered look like parts of a scarified human body. I was awestruck by these clearly man-made objects which were now completely transformed, the glass skeletons of a living sea-skin. Incredibly, and although it had been a familiar sight to the divers and fishermen of Jamaica for thousands of years, this type of sponge was not thoroughly described by scientists until 1987, when the French marine biologists Claude Donadey and Jean Vacelet analysed the collagen structure which apparently gives it its soft and slimy texture. The coat of sponge made the bottles look old, older perhaps than they were. One thing we didn't see were any of the squat brown rum bottles, many of them still corked, which I'd watched videos of Robert Marx and his team handing up out of the water. I was jealous that they'd come across them, picking them up with the first human fingers to hold those bottles in almost 300 years. Bottles containing 'Kill Devil Rum', a Port Royal speciality: a spirit so strong they reckoned Satan himself couldn't handle it. Caribbean buccaneers drank liquor mixed with gunpowder, believing it would give a volatile edge to their characters and make them explosive in combat. Some modern pirates operating off the Horn of Africa drink a cocktail of gin and cocaine in order to ward off evil spirits. The recipe might have changed, but a corsair's regard for health has stayed the same.

The encrusting sponge had preoccupied me with its beauty, which was akin to that of a fallen autumn maple leaf, veined as if somehow human, as if ridged with roots, and burnished gold. The same thing was happening again: it was the life that had claimed the ruins, and not simply the ruins themselves, that begged to be noticed. The ruins were the skeleton and the colonising organisms were new flesh on the bones. I couldn't blame myself for the nature of my attention. It is hard to care how thick a wall is when the wall is thick with life.

I had felt it since I started diving and now I couldn't fight it any more. It was the life – the biological life, the literal life – that made the underwater ruins as interesting as they were, that sustained my passion for them. In the sunken pirate city, a zoological version of Atlantis – a civilisation of sea-organisms – was fused with Plato's drowned human place of punished hubris, and his desolate hazard to boats, his 'shoal of mud'. This was every Atlantis at once. *And people say it doesn't exist*, said my voice of the sea.

Between the two walls of a passageway an arm of coral had grown horizontally, forming an attractive lintel. The philosopher Florence Hetzler described ruins as a collaborative work of art between humans and nature, impossible to produce except by genuine abandonment, and Port Royal was proving her right. Now that I was diving here while pondering her theory, had the theory not become a ruin itself, playing host to new and living thoughts in a mind that swam in the wreckage?

As we drifted on, Llewelyn pointed out the paving stones of the street beneath us, then gestured back ashore. He was saying this road continued on to the land. We were gliding over the remains of Lime Street, a major thoroughfare of the pre-earthquake town. I had walked along a surviving section of it on land and it was spooky to know that this paved road, twenty-five feet below the water, had once been level with the portion that was still up there on the surface. To walk one half of a road and swim over the rest of it is like living in two timelines.

In what had been the streets there were more sea cucumbers, plump and speckled, insouciant of history. 'Over the mirrors meant / To glass the opulent / The sea-worm crawls – grotesque, / slimed, dumb, indifferent,' wrote Thomas Hardy in his 1912 poem 'The Convergence of the Twain (Lines on the Loss of the *Titanic*)'. The sea cucumber may appear grotesque, but when I failed to notice one and touched it, there was neither slime nor indifference: I was shocked at its softness and the way it curled away from the danger. Armourless and exposed, with many natural predators and their only method of defence the sudden ejection of a mess of sticky threads to confuse their assailant, the sea cucumber's every sortie over the drowned roads of Port Royal struck me as courageous. We know that these animals contain basic feelings that also exist in us: yearnings towards food, shelter and favourable conditions, as the marine biologist and 'scuba-diving philosopher' Peter Godfrey-Smith has suggested. The pursuit of a sea cucumber's happiness.

Semi-camouflaged against the mud was the odd conch, a majestic marine snail and the national food of the Bahamas to the north of here on the other side of Cuba. The conch is a china fist full of instinct and muscle. There is something magical about the double-sided character of their shells, triumphantly horned on the outside, but glassy smooth within, the slick living quarters within the fortress where the prized animal dwells. Its eyes are composed of concentric bright rings that peek shyly from the door of its shell, and at other times extend far out on marbled antennae. The queen conch is being overharvested in places. Bahamians who recall simply walking off the beach to pick them up in the shallows are now making expeditions as far as thirty miles offshore to find them.[3]

The ordered paving was suddenly broken by a jumble of bricks. Here there was more wildlife than in the better-preserved sections of the ruins. A school of French grunts, looking festival-ready with their lightning-like streaks of neon yellow and pale blue, were finning gently into the current, looking displeased to see us with their wide eyes and downturned mouths. Beneath them, from two jagged holes in the bricks, a pair of chain moray eels were keeping watch, posed like periscopes. The chain moray has a distinctive pattern of irregular dark-purple blotches and spatters against a yellow background. It works as a very effective camouflage, breaking up its outline against pitted rocks and corals. In blander territory, its patterning becomes declarative: the brash signage of someone who is not to be tangled with.

What can any human know of the interiority of the eel? It is tempting to assume that because eels cannot pull faces like humans and other apes can, there can be no analogous feelings to ours behind their fixed expressions. Meanwhile the eels remain powerless against the thermal and chemical weapons we deploy against them: temperature change and ocean acidification. Microplastics are found in every single seawater sample collected on round-the-world yacht expeditions. The empire of capitalism leaves no drop of the ocean untainted. We must excuse them the odd bite on a scuba diver here and there.

We had been down for over an hour when Llewelyn turned to me and, slowly as though in anticipation of my reluctance, gave the thumbs-up signal. Within a minute we were back at the surface. Our maximum depth had been twenty feet. I'd crossed thousands of miles of ocean to drop just a few fathoms into it, but I'd slaked a lifelong thirst to see the sunken pirate city, the reality that lay behind that impossibly romantic triad of words.

Aflame with excitement, I couldn't wait to ask Llewelyn about what we'd seen. 'I'm sure some of those bottles were seventeenth century! The shapes weren't modern.'

He thought about it, raised an eyebrow and said, 'I wouldn't be so sure. It's possible, but I wouldn't like to say.'

He was right. You couldn't just wildly claim something like that. Still, I could hope.

As Llewelyn and I finned slowly back to the jetty, the profundity of the experience illuminated me like the fierce noon light of the Caribbean. The reason for exploring the sunken remnants of human culture no longer seemed quite so mysterious or difficult to explain. Underwater ruins should make us acknowledge the absurd brevity of much that we hold dear; the fact that the sea reclaims everything in the long run. But the effect they actually have on us seems somewhat different. They look like rare yet magical aberrations, beautiful cemeteries haunted by the ghosts of the legendarily unlucky. Rather than reminders of our mortal condition and the frailty of our creations, they may even work as the opposite: testaments to the stubbornness of a human-made environment. 'Reclaimed from the sea' means temporarily abducted from the control of the sea. We all know this, really. The default state of the Earth's surface is to be covered in salt water and the rest is an aberration, and yet the Antikythera mechanism was eventually found, its rust picked away by analysis, and its purpose, 2,000 years later, once more understood.

It turned out we had slightly more to worry about than the philosophical justifications for diving in the pirate city. Unbeknown to us, as we hovered about in an enchanted state in the mysterious marble-green water, a Jamaican coastguard team were on standby to intercept us for violating the boundaries of the restricted marine area. A coastguard commander had not been aware that Llewelyn and I had been granted a permit to dive. Mr Walters had smoothed the situation over and the coastguard had gone back to their patrols. We had narrowly avoided arrest for an act of underwater piracy.

Along the shore, about twenty men and women were cleaning the morning's catch to be served in the local seafood restaurants for lunch and supper. They sat on the floor moving knives across flesh as though fiddling with musical instruments. The women were humming slow tunes. Nearby, the pelicans and gulls hung around in the hope of some fish guts and heads, and shirtless young boys ran along the short pier where their parents' boats were tethered and, leaping and tucking themselves into cannonball shapes, crashed fearlessly into the water. Port Royal was full of Port Royalians. The old pirate shore was alive.

I went to find Lamoya and thanked her for her encouragement and advice. She smiled and nodded her head.

'See?' she said. 'I told you.'

The Turn of Poseidon

Achaea, Greece

Greece. A winter's night, twenty-four centuries in the past. Five thousand people, many of them sleeping, are alive within the walls of a prosperous city. Under the starlight the chests of men, women and children rise and fall.

The sea laps at the sand of the north beach. The spiders have gathered their webs in the darkness, and dogs with their whining stomachs sleep curled on stone floors, impatient for the coming of morning, the promise of fish-heads and hastily butchered bone. In the fields round about, the sleepless shepherds, as always, hope for a day without loss tomorrow, and the last slave-girls to turn in for their beds are holding their hands to the bases of their spines that under the cool sun have been bowed for far too long.

There is a rumble. The mules in a nobleman's courtyard bray at the north, then flee to the southernmost corner where they keep running in spite of the walls, grazing their heads and chests and knees on the plaster. The earth shakes. Wrath has been kindled in the rock of the earth, in the suddenly restless flesh of the world.

Those who remember the tremors of past days have not known the like of these thunders that roar from the ground. A grandmother cries out, 'Poseidon.' Seizing her son's child from his woven basket, she runs into the wide street where nothing can fall and crush him. *Be still, child*, she thinks, but says nothing. The screams that are scything from the nearby houses would cut down her words.

Something is wrong with the road. The earth bleeds. There are puddles rising, yet there has been no rain. These must be demons of water, or oil, or perhaps this is the blood of Greece itself, streaked with shivers of starlight. The ground is disappearing under the restless fluid black. There are more screams, and more, until there is only one scream, a single wail, a braid of a thousand voices, the cry of a city, loud enough to turn even the heads of the gods.

One god does turn. Poseidon, lord of earthquakes and oceans, has thrust his way from the growling chasm under the sea, and swum to the shore of Locris on the other side of the Gulf. Now, the seawater his muscle, the waves his rippling shoulders, the heavy brine his back, he turns around for Helike, site of his temple, and cradle of those who offended him. In an instant he is upon them. They will know what it is to pray at his bosom now.

The voices are quieted. The dust of the city shimmers on the water. The bubbles of air that break the surface, the last prayers of its inhabitants, finally cease. Satisfied at the destruction, Poseidon has vanished back into tranquillity: he cloaks himself in a raiment of silence. The stars appear still in the darkness. The shining city of Helike is no more.

★

When it comes to the Atlantis story, not everyone fits into one of the two main 'camps' of literalism on the one hand, and vehement anti-literalism on the other. A third set of people read the tale as fiction, but as fiction inspired by actual historical disasters. In their view, these seismic happenings played such a vital role in Plato's invention that they might even deserve to be viewed as the origin of the story: as the 'real Atlantis'. One such event was the drowning of the lost city of Helike in mainland Greece.

By the 1960s, everyone at work in the field of ancient history understood the draw of finding Helike. It was a name that seduced the imaginations of questers to its great promise. If found, the lost city would provide a unique insight into the golden age of ancient Greece: a time capsule of life in the time of the great philosophers. Jacques Cousteau led two expeditions in search of it. Whoever discovered Helike was guaranteed a place in the annals of archaeology, exploration, marine research, and the study of the classical world. It became one of Europe's most elusive archaeological grails, and I was on my way to its environs.

After landing at Kalamata Airport I drove a northerly zigzag up the Peloponnese, arriving in the village of Nikolaiika on the western bank of the Kerinitis River in the late afternoon. This was the locality of ancient Helike, nowadays dominated by a through road along the sides of which lay detached houses interspersed with petrol stations, tobacco stores and striplit truckers' diners. As the moon rose I parked in a gravel yard in the middle of which stood a line of twisted fig trees. It was late October. My hotel appeared deserted. Eventually the owner, an old man with a thick grey moustache and mournful brown eyes, opened the front door and greeted me with the warmth of a worried grandparent. The set of keys he gave me had a dark patina. They looked old and prized.

We might draw a line between those who have searched for an actual Atlantis, and those who have instead sought out somewhere like Helike, a known historical site that may have influenced the story. In its time, Helike had been a renowned and wealthy city. It was a member of the powerful Achaean League, and had a well-known temple dedicated to Poseidon, its patron god. There are several mentions of the city in Homer's *Iliad*,[1] where it is said to have contributed ships to Agamemnon's armada. In the eighth century BC, colonists from Helike founded the city of Sybaris in modern Calabria, Italy. As such, the historicity of Helike has never been doubted. This was a real lost city, one that went from flourishing to annihilated overnight. It happened in the winter of 373 BC, when Plato was in his mid-fifties. Just over a decade later, he would write the *Timaeus* and the *Critias*. What need was there to go searching back in the Bronze Age for a 'real Atlantis', when there was one barely a hundred miles west of Athens that had disappeared within Plato's lifetime?

Few have been more obsessed with the search for Helike than the American scientist Harold 'Doc' Edgerton. Edgerton was not just an academic researcher: he was a brilliant inventor. A pioneer of slow-motion photography, he shot the famous video of the milk-drop coronet, in which a splash of milk creates a perfect, brief white crown of liquid. He also spear-headed the use of sonar to take pictures of the seabed. If anyone could deploy the latest imaging technology to find the remains of a long-lost city, it was surely him.

Paul Kronfield was a young member of Edgerton's crew, a search team champing at the bit to solve an ancient mystery. 'It was the ultimate adventure for a young man fresh out of college, thrust out into a new world of science,' said Kronfield. 'A lot of the people were thinking about the treasures of Helike . . . there's certainly a lot of gold, a lot of valuable things.'[2] Their secret weapon was an instrument patented by Edgerton called a 'sonar fish'. It was a device that looked like a small yellow bomb, with fins to make it hydrodynamic. Dragged behind the boat like prey caught on a line, it sent high-frequency soundwaves down to the seabed. By comparing the delay between the emission of the sound pulse, and its reception by the 'fish' when it bounced back, it was possible to produce an image of the underwater 'terrain'. This, it was hoped, would show outlines of structures – walls perhaps, or the remnants of buildings – or evidence of shifted earth that would help pinpoint the location of Helike.

Old clues told them exactly where to start. The mathematician and philosopher Eratosthenes had visited the area about 150 years after Helike sank. He spoke to local boatmen who told him about the statue of Poseidon that was sometimes visible in the waters of the *poros*, a word interpreted as the 'gulf'. The second-century traveller and geographer Pausanias, famous for his ten-part book, the *Description of Greece*, had noted the position of Helike, the ruins of which could still be seen underwater in his time. Pausanias wrote that:

> *Going on further you come to the river Selinus, and forty stades away from Aegium is a place on the sea called Helice. Here used to be situated a city Helice, where the Ionians had a very holy sanctuary of Heliconian Poseidon . . . But the wrath of Poseidon visited them without delay; an earthquake promptly struck their land and swallowed up . . . the buildings and the very site on which the city stood.*[3]

The submerged ruins of Helike were still being referred to by writers of the Middle Byzantine period, corresponding to the ninth and tenth centuries AD. After this the references dry up. The city became as lost on paper as it had long been lost to the sea.

The city of Aegium still exists – known today as Aigio – and forty stades was equivalent to just under four and a half miles. The location of Helike was therefore clear: it must have lain just under four and a half miles due east of Aigio in the waters of the Gulf of Corinth. This 'Atlantis' was now beneath the sea that once lapped at its coast, waiting for explorers equipped with the latest technology to find its hidden riches.

Cousteau undertook two expeditions in the Gulf. On one of them his vessel, the *Calypso*, was beset by bad weather, flung about in seas uncharacteristically choppy for the Achaean coast. The frustrated searches only made 'Doc' Edgerton more obsessed. For nine years he scoured the seabed for any hint of ruins, systematically combing every square yard within the area mentioned by Pausanias.

In one survey, in water about 130 feet deep, the sonar showed images of 'pockmarks': patterns of shallow holes, arranged in lines up to a thousand yards long. Surely these couldn't be random natural features. Edgerton believed the location of these marks on the seabed of the Gulf, in the position they had found them, could only mean one thing. They had discovered the remains of Helike.

The crew informed the Greek government. A team was dispatched, equipped with an enormous drill on a floating rig. They would pierce the seafloor in the area where Edgerton's sonar had found the rows of marks and, surely, find evidence of the lost city that lay behind Plato's Atlantis.

Years of intensive drilling around Edgerton's pockmarks returned little of archaeological interest and no clues whatsoever as to the location of Helike, besides that it wasn't where they'd been looking for it. After close to a decade gazing painstakingly into the Gulf of Corinth, the lost city remained lost. The mystery lived on.

Just before sunset, I walked the mile and a half down the gentle slope to the shores of the Gulf. It was a still evening and the water was lake-like. There was one small boat a hundred yards or so offshore, and from it two men were silently dipping a fishing net. I crouched by the waterline. Through it the pebbles were visible with a mesmerising clarity, the water seeming to transfigure rather than obscure them. Its restlessness was an interpretive lens that made the stones shiver and twitch with life. I reached my hands down and let them be filled by the old Achaean sea, and lifted it up to taste it. The English swimmer and writer Charles Sprawson quotes Lord Byron as saying, after swimming in the Gulf, that 'there is a unique rapture about a Greek bathe. The mystery of ancient Greece unfolds itself.' Sprawson himself chased this feeling. He wrote that during swims in the storied waters of classical Europe and North Africa, he sometimes experienced 'a momentary lifting of the veil, some mysterious insight into the secrets of the ancient world'.

*

By the beginning of the twenty-first century, Edgerton's and Cousteau's quests had long been abandoned. Dora Katsonopoulou was a professional archaeologist who had been researching the city of Helike for almost twenty years. As a native Achaean, Katsonopoulou knew the Gulf well: it was her home sea after all. As a Greek speaker, she also had a different insight into the words of Eratosthenes. When he wrote, in the third century BC, that the remains of Helike were submerged in the *poros*, he did not bother to explain what exactly he meant by that word, because it would have been clear to anyone reading his work at the time.

Over the centuries, however, the meaning of words can change. Katsonopoulou knew that in antiquity, the word *poros* had referred to a narrow passage of water. This did not necessarily imply the sea itself: it could also mean a stretch of seawater separated from the main sea by a strip of land. She wondered if *poros* did not in fact refer to the Gulf of Corinth, but to an inland lagoon? If so, it would explain why nobody had been able to find Helike: they were looking in the wrong place. Katsonopoulou knew that the local rivers were constantly changing their courses: there were little bridges in the area that had once vaulted streams but now sat over dry ground. The deposition of river-borne sediment carried down on to the plain from the mountains would, over more than 2,000 years, have eventually covered the remains of the city, and this would be consistent with the account of Pausanias, who described how the ruins of Helike had become partially cloaked with silt by the time he visited them in the second century AD. Katsonopoulou realised that if Helike had sunk into a lagoon, then it would not be under the sea, but under the land. All those quests to the bottom of the sea would turn out to have been misguided.

Katsonopoulou and her colleagues believed that Plato's Atlantis story couldn't be disregarded as a source of information about Helike. They had one particular reason. Describing the destruction of Atlantis in the *Timaeus*, Plato wrote that

> there occurred violent earthquakes and floods; and in a single day and night of misfortune all your warlike men in a body sank into the earth, and the island of Atlantis in like manner disappeared in the depths of the sea.[4]

This begged the question: why this 'double destruction'? If a writer were simply making the story up out of thin air, would they not say that the place was destroyed by an earthquake or by floods? Why had Plato used two images? It sounds like an oxymoron, a city sinking into both the earth *and* the sea. Katsonopoulou did not believe this was a coincidence. Perhaps behind Plato's description there lay an eyewitness account of a complex natural disaster. Something thrillingly simple was possible. What if the 'fall of Atlantis' was not the result of a post-glacial meltwater pulse, or a volcanic eruption, but an event which actually fitted with the description Plato gave, of earthquakes and water specifically, in that order? And what if it had happened during his lifetime, not far to the west of Athens?

How could an earthquake make an inland city sink underwater? The answer is a rare but well-known phenomenon that can occur during seismic activity: liquefaction. As the earthquake unsettled the saturated soil below ground level, the buildings of Helike would have begun to sink as pools of liquid materialised like hallucinations in the streets. Two thousand years later, the same thing would happen at Port

Royal. Geological investigations of the cliffs to the south of Helike made it clear that there had been a massive slippage of land at the time of the earthquake: the level of the city had dropped as much as twenty-three feet. The cataclysm, however, did not stop there. Helike was then subject to a second, connected disaster. The earthquake had sent a seismic wave – a tsunami – surging across the Gulf of Corinth towards its northern shore. When it got there it ricocheted off like an echo and raced back across the water, drowning the city. Helike had been destroyed by an earthquake and a flood, just like Atlantis; or rather, the fall of Atlantis had been inspired by the destruction of Helike.

Where would Katsonopoulou and her team start their search for the lost city? It was far from obvious. For two years they drilled boreholes across the area in the hope of finding evidence of human habitation: potsherds – fragments of pottery with specific periods of manufacture – or other small artefacts or datable bones. Though they were speckled with interesting material, the cores they brought up yielded no evidence of a settlement that matched Helike.

They went back to their sources. Pausanias described the ruins of Helike lying forty stades from Aegium (modern Aigio), but that wasn't all he said. He also claimed they were situated in a direct line thirty stades, or around three and a half miles, from a shrine called the Cave of Heracles.

At the foot of the hills behind the plain there was indeed a cave which locals referred to by exactly this name. No one could be sure whether it was the same cave mentioned by Pausanias, a place where the people of Helike left offerings to the hero and consulted his oracle. Katsonopoulou and her collaborators, who included the scientist Steven Soter, decided to err on the side of optimism. They got out their compasses and drew two curved lines on the map, creating two semicircles, one centred on the town of Aigio and extending out the equivalent of four and a third miles, and another centred on the Cave of Heracles, reaching out the equivalent of three and a half. Where the two semicircles overlapped would be their search area. This was the place to dig. If they found Helike there, it would lie not 'beyond the Pillars of Heracles', the location of the fictional Atlantis, but beyond the Cave of Heracles. The presence of the hero's name in both stories is probably coincidental, but it surely raised goosebumps on the two scientists' arms.

It was in the middle of this small ovaloid of land that the team found the ruins of Helike. Katsonopoulou had listened to Plato, to Eratosthenes and Pausanias; she had heeded old shepherds' tales and her own suspicions about an unassuming Greek word. The locals' folk knowledge of the Cave of Heracles was accurate. It had survived since before the destruction of Helike: it was a story older than Atlantis.

★

So, you've found a lost ancient city. How do you prove it's the one you were looking for?

Ostracods are tiny waterborne crustaceans, found in most of the waters on Earth, both salt and fresh. Known as 'seed shrimp' for their size and shape, most male ostracods have two penises, from which they emit coiled spermatozoa up to six times their body length. The oldest penis ever discovered on Earth belonged to an ostrocod: found in Brazil in 2002, it was 100 million years old, and smaller than a grain of rice. Some ostracods breed without sexual contact: in a confluence of myth and science, this is called parthenogenesis, from the Greek for 'virgin birth'. The name 'ostracod' also has a Greek root: *óstrakon* means 'shell', and once upon a time, it also meant 'tile', as in the tile of a roof. They are the most common arthropods in the fossil record and, crucially for Katsonopoulou and her team, the fossilised shells are easily distinguished, with freshwater ostracods having smoother shells and marine kinds having rougher, pitted ones.

Not all the ruins that Katsonopoulou's team discovered were from the same era as the destruction of Helike. Some were far older. The region has a long history of human habitation and they were looking specifically for the remains of buildings from the fourth century BC and the preceding centuries. But the older ruins they found lacked something that was present only in the remains of the lost, drowned city. These were the fossils of marine ostracods. The city had not just been submerged; it had been submerged by seawater. It was Helike. There was no longer any doubt. The 'seed shrimp' helped solve the mystery of the place that had seeded Atlantis.

These diminutive animals were all over the remnants of the city. Its manufactured tiles were themselves clad in tiles of a different kind – tiny, living ones – sparkling with the heraldry of Poseidon on a microscopic scale. This single image symbolised the union of all the Atlantises: zoological, literal, mythological and allegorical. The life of the sea, the inhuman civilisation of the deep, had coated the real-world city that worshipped Poseidon and had inspired Plato's Atlantis. There was even a hint of Jules Verne about the hubris of men like Edgerton and Cousteau with their high-tech sonar. By searching so meticulously for Helike in the sea, and thereby helping to prove that it wasn't to be found there after all, they had unwittingly helped pave the way for the local, Greek archaeologist who succeeded where they had failed. Many-layered visions converged on the place where the whole thing started. It was time to go and have a look.

*

The excavated ruins of Helike lay down a dusty track off the old prefectural road that runs between Nikolaiika and the Selinountas River to the west. I walked down the track for a quarter of a mile until it was blocked by a white gate of painted aluminium. There was no way around. I went to inspect the gate, hoping someone might appear to let me through. When I got there I saw what couldn't be seen from further away: that the path veered left at a ninety-degree angle, then turned back south in the right direction. It was a portal that lay hidden until you had approached it. Sometimes it pays to be suspicious of first impressions.

I was far from the main road now. The scene grew hushed. The air was still and there was little to hear besides the crickets. I walked on, hearing the rocky ground crunching underfoot, and taking care not to be too quiet, which might cause me to be mistaken for a thief. *Keep your wits about you*, said Dad.

I followed the path until it came to a run-down compound. In one corner sat a near-derelict house. Gas bottles and plastic jerry cans lay around. Suddenly the quiet was blown open by a storm of barking dogs. Three of them, an Alsatian and two ragged crossbreeds, roared and snarled as they smashed themselves into the chicken-wire fence that kept them back from me. The Alsatian was having a vigorous go at getting over the barrier. He and the others were chained at the necks, and as they raged their leashes whipped at each other's faces and entangled their feet, tripping them. When the shock wore off I tried to calm them down, holding my hands palms-out in front of me and saying, 'Shhhhh.' It didn't work, so I walked on until the barking faded away behind me.

In the distance the lilac mountains of Phocis stood beyond the Gulf of Corinth. The chirruping of insects merged into a single sound, their many clicks melting into a spill of noise. All around were lemon groves and ancient olive trees with whorled knots on their trunks, slow whirlpools. They looked so old. How ancient, then, is Helike? In terms of the age of the Earth, of the cosmos, it was built only yesterday; but those grand timescales cannot deprive it of its antiquity. Swallowed by the sea in the lifetime of Plato; an underwater tourist attraction for citizens of ancient Rome; lost to history; lost to the sea that had taken it; lost to all human awareness under the land. At last it was rediscovered, and now it is laid out under the sun once again, with seeds drifting into its crannies on the breeze. No place can have been through all those things and not be old and special. If the universe were ten times older than it is, Helike would still be ancient.

In spite of this, the unearthed zone looked freshly vacated. In one sense it had been: archaeologists worked here, talking and eating and thinking and drinking. There was something else, though: it looked like people hadn't long called it home. The block-work was still level. The floors were astonishingly well preserved. An area that had been paved cobble-fashion looked perfectly flat, as though recently laid in place. Homer said that men from this city, *and all the coast-land round about Helice*, sailed east in their ships with the fleet of Agamemnon and fought in the Siege of Troy. Was I staring at the cobbles where they had walked? Had Katsonopoulou and her team touched them, their hands the first to feel the stone in well over 2,000 years? No, it turns out: this building was a dye-works from the Hellenistic period, or the last three centuries BC. It was too recent for people of the Bronze Age, traditionally dated to between 3300 and around 900 BC, to have lived and worked here.

Some objects looked new: shockingly so. A cistern was smoothly plastered inside, like a shop-fresh bird bath. It could hold clean water, now, today: see a family through dry times. At that moment it held only air and light, and it shone with an intensity matching that of my disbelief at its condition. The city had been wiped out overnight, spent thousands of years under water and mud, and still been discovered in an excellent state of repair. The sea had not just destroyed, it had preserved. The silt and soil had come in afterwards and done the same job using different materials. Destruction had, in a sense, been kind to Helike.

Akrotiri, Baia, Port Royal and ultimately Plato had all led me to this place. I was on dry land but I was diving, through the absent soil and the longer-absent water; diving through time. The earth cannot be leapt into like water, only shovelled and chipped and brushed away, and looked at afresh as Katsonopoulou had brought new vision to the words that led her here. Maybe this should be the motto of the Atlantis-seeker: LOOK BENEATH. Consider. Who knows what you will find?

★

When I got back to my room I searched through online academic journals to find more articles by Katsonopoulou. One was an essay called 'Helike and Her Territory in Historical Times', in which she used a combination of pottery analysis, knowledge of contemporary texts, and her familiarity with the geography of the area – the actual lie of the land – to speculate as to the possible boundaries of the lost city. Katsonopoulou observed that 'the site of Helike . . . may have included on the East the area occupied today by the modern villages of Rodia, Elaion and Zachloritika. Toward the West, it extended as far as the Selinous River, which separated Helike's territory from that of neighbouring Aigio.'

I drew red lines on the map representing these limits. They enclosed an area almost seven miles wide and stretching several miles inland from the coast. When I looked at what was in front of me, I sat there astonished, staring at the space between the red lines. The excavated portion of the city must be a comically tiny part, the equivalent of a postage stamp in a garden. The ancient city and its suburbs covered eight to ten square miles. My hotel was right in the middle of it. The truckers' café, the petrol station, the diner in which I was the only customer, eating souvlaki with a family as they played with their child: all of them fell within the likely limits of the ancient city. No wonder it had managed to found overseas colonies of its own, played a key part in the Achaean League, been mentioned in the *Iliad*. I thought I had been lucky to spend an hour at the site. I had actually spent my whole stay there, eating and sleeping above the hidden remains of Helike. While walking on the tarmac pavements of modern Greece, I had been hovering over Atlantis, as a diver floats over the seabed.

Katsonopoulou's discovery of Helike proved something besides where the long-lost city was located. It proved that you don't always need millions of dollars and a ship to make important discoveries. You don't need computers or algorithms. You can have breakthroughs by way of focussed thinking alone. Katsonopoulou had considered her way to Helike. The pickaxes and trowels had followed her insight, until they struck what had become myth and turned it back into reality.

The fact that Helike may have inspired the creation of the most famous 'lost city', Atlantis, added an extra layer of lustre to its name. Nor does the story end there. Beneath all this, beneath classical Helike and the Atlantis story, Katsonopoulou and her colleagues found yet another layer: the Early Helladic Bronze Age settlement, itself 'destroyed and submerged by an earthquake . . . as happened to its Classical successor some two thousand years later'.[5] It was layers and layers of fascination, of reality underpinning story, and story inspiring the quest to dig into reality. It all supported my belief that Atlantis is the most interesting place-name in the world: the name of a place both real and unreal.

★

The twentieth century. A young woman stands on the edge of the Gulf of Corinth, staring into the calm blue of the ancient sea. Her thoughts go to two lost cities, Atlantis and Helike, and what it would mean to find them. Many men had come looking: men with binoculars, men with diving equipment and gadgets, men with ships. But she is an archaeologist, she has proven her credentials, and this is her country. She is an Achaean, a native of the very coastline from which Helike disappeared. If its ruins had been visible once then they could be found again. There must be a secret key, a key that through study and intellectual application might be discovered. One day she would track down the key to Helike.

He Walketh Through Dry Places

Pavlopetri, Greece

The Malea Peninsula lies just to the north of a glimmering strait of water. I arrived there just before sunset and looked across to the island of Elafonisos, a boomerang of rock and shrub with a small port town on its leading edge. It was a swimmable distance for someone bolder than me. The white buildings of the harbour were lit hot pink in the westering sun. I thought about making a dash for it, then realised I would emerge on the other side without a towel and, at this time of day, no means of getting home other than to turn around and swim back. My better sense kicked in. I had no clue about the currents here which, given the narrow channel, might be fearsome in places. *Sprawson would have just done it*, I thought. *Byron would have just done it.* I banished the thought and, a little crestfallen, went to scope out Pavlopetri.

On the way there I thought about how ignorance provides fertile ground for imagination. The classicist and ancient historian Christopher Gill has said that Plato could not have been writing historically about what was happening 9,000 years before his time, because the Greeks 'had no idea what was going on 9,000 years before' his time.[1] Nor did the Egyptians, whose vaunted antiquity Plato used to offset the unlikeliness of his story. The whole thing was set in a time soaked in unknowing. Ancient cultures and oral traditions were full of tales and lore about their deep pasts, but these accounts were something very distinct to anything we would recognise as history. The Sumerian King List, dated to around 2000 BC, makes reference to rulers whose reigns spanned thousands of years. The Greek oral tradition tracked the descent of powerful families back to the gods. Later writers such as Herodotus and Thucydides, who attempted to give sober and reliable accounts of known events, were concerned with the very recent past, within living memory. The haze through which the people of the classical world were forced to contemplate very ancient times was a boon for Plato. He was able to use the unknown, the fog of lost time, as his canvas. As a scientist can use a vacuum to perform powerful experiments, so in the hands of a storyteller, an absence of knowledge can be a resource: a fuel of intrigue.

When I got to Pounta Beach – Point Beach – it was deserted. To the east the ragged peaks of Zovollo and Krithina were fading into the night along with the last hills of the mainland. I knew their names from the work of the archaeologist Anthony Harding, but many places in this corner of Laconia are remote. Internet searches for these isolated high locales return almost nothing. Artificial intelligence has never walked among them.

A breeze blew up from the south and there was a chop on the strait, restless white scuffs on navy blue. Dozens of coloured buoys – green, yellow, red – bounced on the surface.

At around 5,000 years old, Pavlopetri is the oldest known submerged city in the world. Artefacts from several distinct ancient periods have been found there, from the late Neolithic to the Mycenaean. The oldest parts of the walls still visible underwater date to the fourth millennium BC. The span of time that separates the building of Pavlopetri from the destruction of Helike is greater than that which separates Helike from our own time: this would have been an ancient place even to the ancient Greeks.

As I got nearer I was expecting signs warning of restricted access. There were none. There was simply a detailed map of the site, put there by the Ministry of Culture and Sports Ephorate[2] of Underwater Activities. It included some historical information and suggested three routes for exploring the ruins, coded green, yellow and red for easy, intermediate and advanced. So that was what the buoys were for. SNORKELLERS CAN FOLLOW THE BUOYS OR SWIM AROUND, said the sign.

After my experiences in Jamaica, and the state's understandably serious, circumspect attitude about people diving in the ruins of Port Royal, the laid-back approach of the Greek authorities to having random people swimming around Pavlopetri was surreal. I stood there in the lilac dusk, astonished. There was no noise but the slap of the water on the shore, and the restless wind in the thorny bushes behind the beach. *Got the place to yourself*, I thought.

Had I? I looked around by the base of the sign, and further, round about me. Spaced out at regular intervals were several dozen large hollows cut into the rock. They were each between four and five feet long, wider at the top than the base: roughly the shape of Neolithic axe heads. It was a shape that instantly sent a chill down the spine. This was the Bronze Age cemetery, and these were the graves. I had been expecting to see them, but reading about ancient graves is one thing. Standing among them, as the sea darkens and the land gives way to thick and purple shadow, is something else.

There were no bones left to speak of, but this only gave the graves a ghostlier aura: originally they had contained corpses, bodies absent of life, but now they contained space absent of corpses. It was space as a sort of second death. The graves were known as cist-graves, from the Greek *kiste*, meaning a chest or box, and their shapes told the tale of how their inhabitants had been buried, in a foetal position perhaps, or the bones separated and arranged into patterns consonant with customs and belief. Pavlopetri had originally been dated to the Mycenaean period, or around 3,100 to 3,600 years ago, but subsequent research showed the site had been occupied for far longer, with the first signs of human habitation dating back 5,500 years. Two million nights, perhaps, had passed since some of these women and men were laid to rest.

Cut into the orange rock of the shoreline were several larger tombs. One had a small set of steps leading down to a well-preserved square entrance, beyond which lay pitch-darkness and the musty smell of dust that for centuries had felt no rain. My studies of long ago came back to mind unbidden, and I found it impossible not to think of a verse from the Bible, Matthew 12:43: *When the unclean spirit is gone out of a man, he walketh through dry places, seeking rest, and findeth none.*

The cemetery ran along the sea's edge. Many of the graves were completely underwater, and some of them, just on the tideline, were kept full of water by the odd wave here and there. Others were filled with dried *Posidonia* weed and jumbles of rock, bleak offerings of those most consistent graveside mourners: the earth, the wind, and the sea.

My first intention had been to explore the submerged town, but the tombs could not be ignored. They were old, so old that the names of those buried in them had long been washed away from any memory; but they were still here, tenacious in their emptiness, warning of the fragility of human settlements next to the creeping seas of change, but also whispering of our endurance. In some places the sea is neither our neighbour nor our destroyer, but the fellow inhabitant of our tombs. As the sea lapped at the graves of Pavlopetri it seemed to sing in solidarity with our plight. In the meantime, every wave washes away a little more from the rock. The graves are getting shallower. One day their emptiness will simply be an invisible part of the world.

The water possesses a threefold cleansing power: to rinse dirt from the skin, preoccupations from the mind, and the clutter of quotidian concerns from human ruins. In water, all terrestrial things are stripped of something. The sea can play the liquid piranha, whose rasping kiss has the limitless appetite of a god. Even death is continually sanded harder and whiter: as the whale corpse turns into bleached carcass, then skeleton, finally into dust and sand and the tiny clusters of atoms from which the future will be built.

Shaking with excitement from my first glances at a Bronze Age sunken city, I ran back to my little apartment and talked to myself about the beauty of the site while I got in the shower. Then, reaching for the soap, I knocked the tap with my arm and it snapped clean off. Water hissed and squirted everywhere. Within a minute the floor of the room was covered.

The hot water spattered on the tiles, fizzing away uselessly as I tried in vain to get the faucet back on. I winced at the waste in this wind-raw and arid environment. I gave up and went outside, hollering across the courtyard for help. The young son of the proprietors, who wore a slightly embarrassed smile even though it was I who had ruined his evening, turned the stopcock off outside the building and told me the proprietor, his dad, would be back tomorrow to fix the leak.

As promised, the father came round the next morning, apologising for the delay, even though he was early. He got to work but struggled with his assortment of wrenches and cursed at the plumbing, each time peeping out of the doorway to say, 'Mr Damian! Sorry I don't swear at you.' Eventually the shower was fixed, the water at bay, and the proprietor at peace.

★

That night I gathered my snorkelling gear and went back to the beach. As I entered the water I turned my camera on and held it against my torch in my right hand, sweeping the shaft of light from my shoulder to keep it steady. A tiny translucent pink jellyfish entered the torch beam, then a fidgeting shrimp, then a few small squids. I followed the line of the shore while moving out into slightly deeper water until I came to the hollows of the submerged cist-tombs. Fish were loitering in them, hovering ghost-like in scoops of shadow. I continued swimming and saw that there was a shallow grave below me. During the day this had seemed OK: sunlit snorkelling in an archaeological site. By night it felt very different. I imagined the locals thinking I was doing something ill-omened. I changed course and headed back to where I'd got in. There was a concrete block with a rope attached: *ah, good*, I thought, it was one of the marker buoys. But behind it something was moving, a ribbon of darkness. I moved my torch beam to highlight it fully: a moray eel on the hunt. It was a big one, as long as me, its body thick and muscular. With a purity of liquid menace it started to move in my direction. Its mouth was open, giving its wide, round eyes a demonic look. I began to breathe more quickly, turning straight around and swimming rapidly back to the shore. I decided not to get in the water again.

*

Pavlopetri had been 'discovered' by the British archaeologist Nicholas Flemming in 1967, but it seemed absurd to think that locals – fishermen, children, anyone who came down here and swam about in the shallows – were not well aware of the existence and scale of the place. The exposed remains of the town cover almost 50,000 square yards: an area about the size of seven American football fields. Much more of it remains obscured from view, buried under the sands of the Elafonisos Strait. What remains visible, I was to discover the next morning, is breathtaking enough.

For three days I snorkelled alone in the ruins of Pavlopetri. By day, the remains of the city were licked by restless golden beams of light, as though the sun were a disco ball. Although more eroded than the ruins of Port Royal, which is to be expected given their vastly greater age, the walls of Pavlopetri are nearly as distinct as those of the sunken pirate city. This is because of the nature of the seabed and the resultant clarity of the water: there is no mud, and the visibility often extends to dozens of yards. The autumnal water was warm: twenty-five centigrade or so, allowing me to swim for an hour or more at a time. This is almost long enough to forget you are an animal which gets about by any means other than swimming; for the body to forget it was made to walk.

Much of the site consists of straight walls a single course high. In places this can create the impression that it is not the remains of an ancient site, but the outline of a planned one: as though lines of bricks were ceremonially laid out for the construction of a new underwater town. It is an illusion that fades on closer inspection. The stone blocks looked unmistakably old, blemished by the water and the slow travails of molluscs and urchins. Yet as stones often do to hungry eyes, they also resembled food: flatbreads, uneven lumps of dough, or whole cheeses with dirty rinds. I had been swimming a long time and wanted calories.

Other walls led to rooms, chambers, with smashed and sea-bitten walls several courses high. Further out in the channel, close to the tiny islet of Pavlopetri after which the whole site was named, I swam down a narrow passage and came to a T-junction with two large rooms on either side. These were proper spaces which could be dived into and swum round: little cubes of sea. Now they sheltered their spiny residents from the current as they had once protected people from the wind. Their floors were covered in rubble and the brine-bleached remains of echinoderms. I dived down for shells and urchin-tests. Looking up, the frothing surface of the sea made a turbulent ceiling. Whenever I disturbed the sandy floor the little fish would dash in to browse among the billowing sediment.

The few precious times I believe I have managed to keep still and relaxed enough around fish that seemed (to me) to believe there were no real dangers around, they behaved in a way not dissimilar to that of curious puppies or kittens. In the shallows of the Aegean I have watched as initially frightened fish – particularly young saddled sea bream, *Oblada melanura*, with their pearly silver bodies and distinctive black spot by the tail – became slowly accustomed to my presence, and drastically changed their behaviour. Pavlopetri was a perfect place to observe this phenomenon. The walls acted as 'wind breaks' against the current, meaning I could stay still in the warm water, and the snorkel allowed me to keep my face underwater and breathe without emitting bubbles. Over the course of half an hour the fish went from fleeing my every move to coming in and investigating the subtlest movements of my fingertips. *Probably think they're food or something*, the cynical voice in me said, before I remembered that I, a human, had just now looked at rocks and salivated for bread and cheese. Nor was I necessarily right about the fish. Long after discovering that my fingers were not something that they wanted to eat, the bream kept on touching them, trying to work them out. They were puzzling over them, weighing them up in their minds. One fish in particular swam slowly and repeatedly up to my face and then back to my fingertips as though investigating a possible connection between the two: *my self*, I thought. Was I anthropomorphising? I had come to dislike that word, with its soft implication that feeling connected to anything non-human and alive is intellectually irresponsible. Was the sea bream ichthyomorphising — projecting its notions of fish-ness on to me? I cared less about any of this than about keeping quiet and looking another living thing in the eye, an experience that is not question or answer, but somehow neither and both at once.

Pavlopetri is under threat from several angles. The hulls of passing boats, the ballast water released by larger ships, and the waste discharged from the ferries moored in nearby Vatika Bay are all harmful to the site in different ways. Because the remains lie in such shallow water, a boat's keel can easily scuff and score them. Ships' waste effluent alters the pH of the water, which can eat away at the stone, and ballast water contains solid particulate and micro-organisms that change the ecology of the local seawater: alien sand and alien life. It is an 'invasion' both inanimate and animate, posing a two-pronged waterborne threat to the city's remains. This ghost city, reclaimed thousands of years ago by the sea, is now subject to a second phase of destruction: one that threatens not human lives but a history, a photographic spectacle and a home for animals.

Sometimes my attention would drift from the ruins and fish until it was present only in the slow movements of my breaststroke and the perception of winking sunlight and the caress of the water. Part of the appeal of swimming is very simple. At the end of a stroke, there is a moment when your body is still, yet continues to glide forwards; when you keep moving after you have stopped moving. The human body pushing through the water, whether quickly or slowly, is experiencing something else: the bliss of hydrodynamics. This is the moment when the same water that had resisted the start of your stroke seems to change its mind and help you flow onward. It has been persuaded, stroked into a state of complicity.

This is why swimming is attractive, for some addictive, and for others an obsession. It is reminiscent of the human journey, our bodies labouring in a world which both resists and aids it, drags on it and supports it, pushes it yet then invites it. It is reminiscent of dance; of physical negotiation. Alongside the breath, the chill, the surging power of the sea, there is a kind of spirit that denies you until it applauds your persistence; cold shoulders giving way to open arms; scorn turning into love.

The divers who had gently resisted me at first had become my friends. Atlantis, in all its forbidding complexity, its mystery and age, had opened its doors of meaning at last. But only because I had set out at all in the first place. The stroke can only lead on to the journey if at first, though alone and full of doubt and fear, you reach out.

Future Atlantises

The Coasts of a Rising Ocean

There are few fears more elemental than that of invasion by the sea. HOUSES FALL INTO THE SEA is the headline that sums it all up: the edges of the country crumbling away, their inhabitants sent scampering inland to a place that's just as equally doomed in the long run of time. On a waterfront in my home town, a large sign speaks with unpunctuated urgency of DANGER SUBMERGED SEA DEFENCES. It is a sinister warning, working on three levels: firstly, we need to defend ourselves from the sea, an enemy that is always at the gate; secondly, some of these defences are submerged, which suggests we are losing; and thirdly, when our defences are lost, even they become a new hazard.

I realised, having grown up by the sea, why the locals called it 'the seafront'. 'The beach' made it sound like a holiday venue, which wasn't right for a place that was so often forbidding. 'The shore' made it sound too remote. In 'the seafront' was the truth of the place. I had thought about the other, military sense of this word before, but time spent in the water had tweaked the sense of it. It was a 'front' in the sense of a place where forces encountered each other. It told, in water and stone, of a frontier, the unconquered country of the sea.

The sea is presented as perennial aggressor; its victim, the nation, as fighting an eternal rearguard action against an adversary so powerful it can fling our biggest ships about like corks. Schoolchildren laugh when told about the emperor Caligula, popularly depicted declaring war on the sea and ordering his men to collect chestfuls of shells as their spoils of battle. But the language of news reports has a dourer tone. In February 2014, 'wave action' – note the military tone – washed away a section of coastal railway and its protective sea wall at Dawlish Warren in England. Helicopter footage depicted the nonchalant brush and draw of the water sucking the shingles and rocks from beneath the rails, as if pulling the rug out from under human industry. The imbalance of power was laid bare: the water versus the warren. Poseidon against the rabbits.

Akrotiri, Port Royal and Helike had something in common: like the Atlantis of Plato's story, they were swiftly devastated. Whether or not we believe there was a divine hand in it, their fates were wrought quickly, in spans of hours, by calamitous forces of nature. Hasankeyf, Lion City, and the SS *Fortuna* – two cities and a boat – shared something else. They were all sunk by humans on purpose: the ancient towns by dams, and the modern ship by a floating mine.

Baia and Pavlopetri, by contrast, were 'slow Atlantises' – places submerged over years, over decades, centuries even. These slow immersions were natural, but would we describe them as natural *disasters*? The word 'disaster' has connotations of the sudden, the surprising, the unforeseen. The emergency. In order to see something as a disaster, humans seem to require that it must be abrupt. If it happens more slowly, we are prone to seeing it as natural, and as a gradual shift in circumstance to which we can adapt. Is evolutionary change, for example, a disaster? Was the end of the last ice age? Is climate change?

Global sea levels are rising rapidly. In the nineteenth century their level shifted a little each year but was relatively stable, remaining at more than 100 millimetres below its level in 1980. Since then, global sea levels have risen almost 150 millimetres. The rate of sea level increase went up from 1.4 millimetres per year throughout most of the twentieth century, to 3.6 millimetres per year in the decade leading up to 2015.[1] On the coastlines of the continental United States, high-tide flooding is now 300 per cent to 900 per cent more prevalent than it was five decades ago,[2] a serious and potentially devastating environmental concern for the largest economy in the world. Previous events, notably the local and central government responses to Hurricane Katrina in 2005, illustrate that even countries well equipped to deal with environmental emergencies do not always do so as competently as might be expected. Political factors, expediency, the socio-economic or ethnic status of the likely victims: all these things play a role in the human response to disaster. Where there is danger and suffering, inequality prevails: entrenched and exacerbated.

Sea levels do not rise uniformly everywhere on Earth for several reasons. The system of ocean currents and regional variations in the strengths and directions of winds result in the sea 'bulging' in certain parts of the world, and in these places, sea levels have gone up by as much as 200 millimetres since the dawn of satellite photography.[3] To look at a map of these changes is startling. Areas experiencing the most rapid rate of rises include the north-western, central western and south-western parts of the Pacific Ocean, the western South Atlantic and Southern Ocean, and the western North Atlantic. Asian countries will bear the brunt of this change, with China, Vietnam, Japan, India, Bangladesh, Indonesia, Thailand and the Philippines home to the largest numbers of people likely to be affected. In China alone, more than 40 million people live in threatened coastal locations and are in danger of being displaced by a swollen ocean. In Africa, the densely populated delta regions of Egypt and Nigeria are especially vulnerable to water rises. This does not mean wealthier nations can simply breathe a sigh of relief. Almost half the population of the Netherlands lives in areas at risk from rising seas, and in the United Kingdom, 3 million people live close to shorelines and riversides prey to worsening floods. More importantly, our world is interconnected in every way we can conceive of and doubtless many other ways we do not. The consequences of change in one place are felt, in some manner, everywhere else: demographically, economically, chemically, emotionally.

In addition there are particular 'black spots' of sea level rise located around the world: areas where precise local factors could bring extreme change. From Japan and the Philippines to the Pacific Islands, to South East Asia, the southern tip of the Arabian Peninsula, to the seas around Mauritius and the Gulf of Mexico, there are places where the sea will rise faster and higher than elsewhere.

In some places a little sea rise goes a long way. Countries with extended flat shorelines, with extensive lands which do not rise much above sea level, or indeed lands that lie below sea level, face greater peril as a result of rising seas. Some of them, like the Netherlands, while in serious danger of prolonged marine catastrophes, are at least in an economically and technologically strong position to confront them. Others are not. Some 32 million Bangladeshis live in areas that are extremely vulnerable to elevated sea levels, and there are estimates that by 2100 up to 13.3 million of its citizens will have been forcibly displaced by climate change.[4] The same is true of at least 27 million Indians, and almost all of the 3 million Pacific Islanders who live within ten kilometres of the coast.[5] Baia and Pavlopetri are not just survivals of the past. They are visions of many places' futures. They are predictions.

The Maldives is one of the most dispersed countries in the world. Its 1,190 islands, which are mostly composed of coral sands, comprise only 115 square miles of land area. They sit atop a chain of atolls and their average height above sea level is only one metre. It is the flattest country on Earth. The Union of Concerned Scientists estimates that, 'given mid-level scenarios for global warming emissions', the Maldives could lose 77 per cent of its land area by 2100.[6] Homes and critical infrastructure are densest along its coastlines. The Maldives, as its people know, is unlikely to survive. The island nation of Kiribati, situated in the middle of the Pacific, faces a similar fate. Two thirds of its land will likely be gone by the end of the present century.

Jakarta, the Indonesian capital and home to 10 million people, has been called 'the fastest-sinking city in the world'.[7] The extraction of underground water supplies has caused huge areas of the city to subside by more than five metres since the 1970s. In tandem with this problem, rising sea levels mean Jakarta could be largely or even completely submerged by 2050. This situation is far from unique. 'Dhaka is sinking,' wrote Shamsuddin Illius in January 2023.[8] The giant capital of Bangladesh, the ninth largest city in the world and the seventh most densely populated, is also suffering from the disappearance of its groundwater, which was once struck at depths of twenty to thirty metres, but is now so depleted that engineers have to seek it hundreds of metres underground. The resulting subsidence makes it even more vulnerable to rising sea levels. The same is true of Bangladesh's second city, Chattogram. Around the world, other coastal capitals, including Lagos and Bangkok, face similar futures. New York City will confront increased risks of severe flooding. Whatever the prospects of constructing defences, an inevitable consequence of this will be human migration on a large scale. This is not a simple case of moving inland and uphill. Countries for whom the consequences of climate change are severe will become less viable places to live. People do not simply leave a broken coast and retreat to high ground. Those who are able also take to the roads. They take to the air. They take to the seas.

At the bottom of the Koprinka Reservoir in Bulgaria sit the remains of Seuthopolis, the royal city of King Seuthes III. The architect Zheko Tilev has proposed a spectacular 'un-drowning' project costing tens of millions of euros, which would include the construction of a huge circular dyke around Seuthes's long-drowned city. The dam would be raised high enough for the water to be drained away and the palace made dry once more, a resurrected Atlantis sucked upwards into the modern world. Visitors would take boat trips out to the walled-off 'island', the surface of which would sit beneath the reservoir's water level, and enjoy coffee and sandwiches and the opportunity to buy small high-quality gifts. They could walk along the dam wall, a rampart built to defend a city without a population.

The drowned place becomes a sort of hallowed crime scene with the water cast as villain: a memorial site like a battlefield, meriting reverence and dignified information plaques. Even when our language is less martial, it remains solemn and funereal. We speak of ships' graveyards and the resting place of the *Titanic*. We believe in our right to bury our comrades at sea, but we hope not to be killed by the sea while still on land.

We may think we have better things to do than de-submerging long-lost cities. Coasts are at risk and action, both the sea's and our own, may now begin to outpace hope. 'Most of the Netherlands is already below sea level but is not disappearing, because the Dutch are building and maintaining their coastal defences,' observes Gerd Masselink, professor of geomorphology at the University of Plymouth.[9] Building walls is only one solution, and its limitations include the facts that walls are expensive, made of materials which are globally in limited supply, and require constant human maintenance. Walls are not self-sustaining. They are 'hard' defences. They deflect water rather than absorbing it, meaning the energy it contains is directed towards damaging other parts of the locality. There are other ways besides the 'hard' way, though. Permeable pavements, for instance, allow floodwater to seep through the ground, rather than attempting – and failing – to channel it all into a human-built drainage system.

More superior still are living systems for coping with the impact of water on the coast. Dr Rachel Gittman works in East Carolina University's Integrated Coastal Programs. She is involved in research into 'living shorelines', or how humans can aid natural systems to cope with the sea, often in ways that benefit human societies. What a 'living shoreline' defence looks like will depend on the local environment, its climate, flora and fauna. In many places a salt marsh works as an intermediate absorptive environment which is neither dry land nor fully marine. In other areas, mangrove forests act as a bulwark against weather events and rising water. At the Chesapeake Bay Environmental Center in Maryland, the waterline has been stabilised using materials which are sympathetic to the return of biodiversity.[10] A restored oyster reef now works as a breakwater. Submerged and semi-submerged vegetation holds the saturated soils together. Nothing is better at this job than the roots of a plant. Not only do they maintain themselves, they multiply. They do not get weaker with the passage of time and weather, but stronger. They sequester carbon, the excessive release of which is precisely why they are now increasingly needed. They are the sea defences that build themselves.

In the oyster, too, there lies salvation. Oysters, a keystone species in many coastal regions, have been found to work as a powerful ally in coastal restoration and flood defence. Scientists call them 'eco-engineers'[11] and talk of 'recruited oysters'[12] – they are not just bivalves, they are colleagues. On the mudflats of Kutubdia Island, south-east Bangladesh, where many people have already fled their homes due to encroaching seawater, Mohammed Shah Nawaz Chowdhury, associate professor of marine sciences at the University of Chittagong, has led an oyster project with startling results. Reefs built by seeded oysters have been found to cause significant wave dissipation and to block smaller waves entirely, as well as bringing about a 54 per cent reduction in coastal erosion compared to un-'oystered' sites.[13] The oysters create natural sediment banks which accrete behind the reef structure, not only protecting the integrity of the shoreline, but also filtering the water and encouraging biodiversity by providing a sheltering habitat to invertebrates, crustaceans and fish. 'It's a dynamic process – not hard concrete,' as Aad Smaal, emeritus professor of sustainable shellfish culture at Wageningen University, told the BBC.[14] 'And that's the new understanding of using natural forces to achieve our goals.'

A 'new understanding' it is. Projects like these run counter to a prevalent human tendency: to think that natural challenges call for unnatural responses. Nature only has the power to destroy because in the first instance it had the power to create. Working in concert with nature sounds less bombastic and spectacular than building gigantic concrete walls. These approaches have a terminology of encouragement, observation, tweaking, protection and restoration. They make less reference to human powers because they acknowledge one that is greater.

Not everywhere that is threatened by marine incursions can be saved. Salt marshes, mangrove forests and oyster reefs are options that only work in certain environments, and there are coral islands where the building of sea defences is a physical impossibility. Like the inhabitants of all Atlantises, those of history and those of imagination, we will see what happens. We must help nature and each other if we are to soften the wrath of Poseidon.

Epilogue

Lokroi, Greece

A chilly, grey morning on the coast of central Greece. I stared out across the North Euboean Gulf, its fidgety surface the menacing dark blue of water under cloud. Half a mile away across the strait was a small island dominated by three hills. It is uninhabited except by its ranger, its wild seabirds and raptors, and the majestically horned kri-kri ibex, or Cretan goat. Local hunters pay good money to be ferried across to the island and shoot the goats.

The previous day, I had asked Dimitra, the owner of my hotel, for some help making enquiries with local boatmen to see if I could catch a lift over to the island. After a long phone call to a fisherman known to give lifts, Dimitra put the phone down and turned to me.

'They are asking, what do you want to hunt?'

'Nothing,' I said.

Dimitra's eyes narrowed a little. 'Nothing? So why do you want to go to the island?'

'Well, I suppose I am hunting the source of Atlantis.'

'Atlantis? But you know Atalanti, here, is named after Atalanta. She was a runner. A hunter.'

'But do you think the names could be connected?'

'No. I think not.'

Besides her duties at the hotel, Dimitra was an athlete and physical education teacher. She was keen to make sure that I understood the place was named after a heroine who shared her skills and priorities.

The island to which Dimitra was trying to help me gain passage is known by several names: Atalanti, Atalantonisi, Talandonisi, and these names all have two things in common. Firstly, according to local tradition, they can all be traced back to the legendary Greek heroine Atalanta, the swift-footed huntress who shot the Caledonian boar, the only female member of the courageous Argonauts. And secondly, they all look and sound a bit like Atlantis.

Long ago, this little island had been considered important. By the time of the Peloponnesian War – the struggle for supremacy between the city-states of Athens and Sparta – it was called Atalanta. At the start of the war the Athenians equipped Atalanta to defend the coast of Euboea from seaborne attacks. They built a watchtower on the island, and garrisoned it with a few soldiers. It wasn't much, but it was better than nothing.

The fortifications were short-lived. In his *History of the Peloponnesian War*, Thucydides wrote that in the sixth year of the hostilities, an 'inundation also occurred at Atalanta, the island off the Opuntian Locrian coast, carrying away part of the Athenian fort and wrecking one of two ships which were drawn up on the beach'. He also described how the sea, 'retreating from the then line of coast, returned in a huge wave and invaded a great part of the town, and retreated leaving some of it still under water; so that what was once land is now sea; such of the inhabitants perishing as could not run up to the higher ground in time.'[1]

This happened in 426 BC. Plato was born between 428 and 423 BC. Whether or not he was alive when the tsunami hit Atalanta, he would likely have known about it. Could it have been another inspiration behind the story: an ingredient, alongside Helike? The similarity of the name was too obvious to ignore, but there was also the rarity of these kinds of catastrophes. The seismologist Iain Stewart has written that 'the parallels between Plato's Atlantis and the 373 BC and 426 BC earthquakes are enticing, particularly given that even after two and a half millennia of notable historical seismicity in Greece, these two earthquakes stand out as particularly catastrophic events.'[2]

It has long been acknowledged that Atalanta Island might have inspired Plato's creation of Atlantis, but it also seems ridiculous. The question has always been how the inspiration makes any sense. The American author Lyon Sprague de Camp insisted the two must be connected; that the similarity of the names, and the fact that the fortifications on Atalanta had been destroyed by a tidal wave, was too much of a coincidence to be overlooked. But how could so tiny an island, manned by a skeleton crew of men, have sparked the thought of the greatest seafaring nation that ever existed?

Dora Katsonopoulou had shown that there was much to be gained from contemplating individual words. I thought of the ruins of Helike, and the nearby neighbourhood of Eliki that had such a similar name. Then there was the Cave of Heracles, its name and location preserved for millennia by local tradition. *Atalantonisi*. Maybe the name of the island was trying to tell me more than I realised. It didn't just look like the word 'Atlantis'. It reminded me of something else. Then I remembered that the sounds of Greek vowels have changed over time. Could it once have been not *nisi*, but *nesi*? That would have made its name uncannily similar to *Atlantis nesos* – 'the island of Atlas', Plato's full name for Atlantis.

I looked again at the island across the water. Was its meagre size really a problem? Or was it a part of a riddle, left by a mischievous philosopher? The symmetry of its three hills, a big one in the middle flanked by two smaller ones on either side, was reminiscent of Plato's description of Atlantis: the hill of Cleito at its centre, surrounded by a great moat, with raised land outside it. Of course, every attempt to make a single place map on to Atlantis was doomed, but what if the whole point was that *Atlantis nesos*, at some level, was a purposefully inflated version – an absurd version – of *Atalantonisi*? Had Plato deliberately turned a puny island into a monstrous one? Had he borrowed its name and given it a new etymology – Atlas's island, instead of Atalanta's – as his devotee Tolkien would do centuries later, when he named his own Atlantis *Atalantë*, 'the downfallen one'? Wouldn't it be a witty test of his own abilities, to see if he could alchemise the tiny into the vast? And wouldn't it chasten the hubris of men who had quested after a mighty Atlantis beyond the Pillars of Hercules, if the real source of the name they sought had been a heroine all along; if the Atlantis Casino in Reno, the Atlantis Palm in Dubai, the various films of spandex-clad underwater supermen, had all been named after a legendary woman, by mistake? In the story, Poseidon had abducted the woman Cleito to produce the king of Atlantis. Had Plato abducted Atalanta in order to create his name?

Maybe Dimitra was both right and wrong. No, the names were not connected any more, because Plato had deliberately uncoupled Atlantis from Atalanta, and sent his version adrift on a different stream.

Even if none of this could be proved, it was of lesser importance than the moral of the story. Beyond any resemblances of shape and name, Plato's story was about hubris, after all: the overconfidence in human power that gets its due reward. Even Atlantis, the mighty faraway island empire, with its unassailable walls and canals, its 1,200 ships and countless fighting men, could not stand up to the sentence of the gods. What could be a more potent indictment of imperial hubris than to base a tale of Atlantis the maritime colossus on a puny offshore holm with a single watchtower and a handful of guards? For anyone who read about Atlantis and stalked its name back to its root, the meaning would be clear: whether menacingly great or laughably small, your human schemes can all be smashed by the judgement of Poseidon. If this were Plato's idea all along, then best of all, he was making use of humour, that underrated intellectual tool, and Dad's favourite means of dealing with disaster. Plato's grandiose tale of a terrifying sea-kingdom was inspired by a few men keeping lookout from a glorified shack on the beach. In our time of rising seas, it gave the story a chilling undertone. Push the gods too far and your fortifications, grand or humble, are going to be found wanting.

It was more than a punchline. It felt like the end of the riddle. I had an overwhelming urge to swim to it, see the little silver fish again, and set foot on the real Atlantis: the perfect place from which to laugh across the sea at Plato's wit and, wet from the water, to shiver at the thought of what might be coming for us.

I didn't know if I would make it to the other shore. The current was strong and I was afraid to touch the unknown coastline of the island; worried it might pierce the dream.

As I readied myself to dive in, I thought I heard a voice. Dad? Rán? Or was it me? *Keep your wits about you.*

I thought back to some of the mistakes I'd made: not the accidents, but the ones that seemed a bit reckless or stupid. A broken depth limit, an attention drift that caused me to lose sight of the group, an unannounced venture below. Why had I done these things? And why had I felt a sense of completion when the divemasters had spoken to me afterwards? When they remonstrated with me I felt shame, but I also felt reassured. Frustration built up into a 'diver error', but with the correction came peace.

Did I miss having someone there to correct my course? Did I want someone to tell me off? It was obvious, the truth like an island rising from the sea.

I wanted my dad.

And in the ocean, inside myself, I had found him.

I smiled. Then I hit the water.

Acknowledgements

So many people have assisted me in the writing of this book that I do not know all of their names – in particular, all the divers who gave me a helping hand with my gear, or who pointed things out underwater while we swum together as a large group, flung together by chance and the ocean. I owe all of them a debt of gratitude, alongside everyone I have buddied with on dives.

I thank the wonderful team at Chatto & Windus/Vintage who have helped turn *The Drowned Places* from a chest full of disordered sea-notes into an actual book: Huge thanks go to Clara Farmer for her belief that this book would work; to my brilliant editor Rose Tomaszewska, whose thoughts on structure transformed it for the better; and to everyone who has tirelessly tweaked the book into shape and helped usher it into the world: Chatto's new editorial director Molly Slight, assistant editor Asia Choudhry, managing editor Rhiannon Roy, copy-editor Sarah-Jane Forder, production controller Polly Dorner, and publicist Susie Merry. Joyous thanks go to Matt Broughton for his beautiful cover design. I thank my agents Eve White, Ludo Cinelli and Steven Evans, without whom I'd never have got started on this. I thank John Hewitt for his magical drawings for the chapter heads, and Tara Darby for her wonderful photographs.

My deep thanks go to my family for their love, support and sufferance of wet dive gear, especially Delaine Le Bas, Lincoln Cato, Emma Baars, and my dear late great-grandmother, Julia Jones (26.01.1927–19.08.2022). I thank Ida and Nina for their awe of the sea, which inspires me every time I share in it.

I thank everyone at Ocean View Dive Club, Sussex, for their friendship and enormous help as I started my journey into diving. If anyone reading this is local and has been thinking about scuba or freediving, do get in touch with Ocean View. Thanks especially to Graeme, Angela, Carl, Jen, Nikki, Teresa, Phil, Paul, Mark, Kenny,

Sam, Laura, Gareth, John and Geoff. In particular I thank Stuart Main for getting me off the hook above the wreck of the SS *Fortuna*, and being a great dive buddy in a tricky moment. I owe you one.

I thank the Society of Authors for the awards which helped me undertake the travelling necessary to write this book, and the Jamaica National Heritage Trust for their kind permission to dive in a protected area. I thank Angela, Carl, Llewelyn, Kees, Marc, Carlo, and all the divemasters who guided me through unfamiliar waters.

Many thanks to all my friends who have sent constant messages of encouragement during the long genesis of this book, but especially Simon Bracken, Jon Day, James Garvey, Thomas Gatley, Ludo Hughes, Christopher Jones, William Kraemer, Robert Lomax, Charles Newland, Phillip Osborne, and John Owen.

I thank those who accompanied me on parts of this adventure, and whose life's journey has since borne off on currents of its own. I thank the JSTOR online library of academic journals for their work in making research material available to the general public, free of charge. Without JSTOR, and the access it provided to essays on everything from Greek archaeology to scuba science and the biology of sea slugs, I simply couldn't have written this book.

Any and all mistakes in this book are solely my responsibility.

Notes

The Isle of Atlas: Plato's Atlantis

1. Hercules was the Roman equivalent of the Greek hero Heracles. The name Hercules has been more frequently used in later Western culture.

Atlantis Since Plato: Auschwitz; Byzantium; the 'New World'

1. Tom Holland, speaking on *The Rest Is History* podcast, episode 315, 'Atlantis: Legacy of the Lost Empire', with Dominic Sandbrook, originally released 23 March 2023.
2. There is a school of thought which says there was no such thing as 'a Renaissance' in the sense of a unique, discrete period of rediscovery and adventure in philosophy, art and science in Italy, but the fact that it is even possible to debate whether these centuries deserve to be called a Renaissance suggests there's probably something in it.
3. Thomas Ward, 'Sixteenth-century philosophical and linguistic strategies: mental colonialism, the nation, and Agustín de Zárate's largely forgotten "Historia"', *Latin American Literary Review* (2013), vol. 41, no. 81, p. 27: 'During the sixteenth century, Spanish intellectuals called *letrados* developed as part of a colonisation strategy what I would like to call a discourse of ignorance.'
4. 'Anti-imperial' does not imply 'pro the masses'. Various commentators have pointed out Plato's discomfort with the working-class rabble-rousers of the Piraeus Harbour district west of Athens. They had played a central role at the Battle of Salamis in 480 BC, but in the Atlantis story, Plato heaps more praise on the land forces, who in his day were mostly derived, like Plato himself, from the upper classes who could afford their own armour and weapons.

The Puzzle: Marburg, Germany

1. Eric Otto Winstedt, *The Christian Topography of Cosmas Indicopleustès*, Cambridge University Press, Cambridge, 1909, introduction, p. 8.
2. Being a giant, Blake's Albion has been associated with Atlas the Titan, but by making him a son of Poseidon, Blake may have been deliberately linking Albion to the kings of Atlantis, who were said by Plato to be the offspring of the god.
3. Though this is based on a false etymology of Britain which can be traced to Isidore, Bishop of Seville in Spain in the seventh century AD.
4. Schliemann's excavation techniques involved clearing out layers of debris in the hope of revealing walls and treasure, thereby destroying enormous amounts of evidence including datable potsherds, bone fragments and identifiable strata of accumulated detritus. His practices have long been shunned by professional archaeologists seeking to approach ancient sites in a more sensitive, productive, painstaking way.

New Questers: The Straits of the Mediterranean; the Deep Sea

1. It is now understood that the Philippine flying lemur, *Cynocephalus volans* or the *kagwang* in the Filipino language, is not actually a kind of lemur at all, but is monotypic, i.e. the only species in its genus.
2. Not everything Donnelly said about Atlantis has been quite so well remembered. He also thought it was the ancestral home of red-haired Irish people, and that the reason there were bananas in both Africa and the Americas was because they were farmed on Atlantis.
3. Robert Graves, *The Greek Myths*, Penguin Books, Harmondsworth, 1955, p. 146.
4. Kaplan was speaking during an episode of the BBC Radio 4 programme *In Our Time* on the subject of infinity (originally broadcast on 23 October 2003).
5. Graham Hancock, *Underworld: Flooded Kingdoms of the Ice Age*, Penguin Books, London, 2003, p. 63.

6 Such as a large spread in the *Minneapolis Sunday Tribune* in October 1931, which included a picture of a 'Rare Armor-Clad Figure Found in a Grave in Mexico and Believed to Be of Atlantean Origin', a statement for which there was no evidence whatsoever.
7 Peter Popham, 'Architect's mission in Cyprus: one man's quest to find Atlantis', *Independent*, 24 August 2005.
8 'Lost city of Atlantis "found in Mediterranean"', *Sydney Morning Herald*, 16 November 2004.
9 *Atlantis Rising*, dir. Simcha Jacobovici, distributed by National Geographic, original release 29 January 2017.

Hermits: Widewater Lagoon, England

1 The name means 'the Cambridge star of Fezouata', as it was discovered in the Fezouata Shale of the Anti-Atlas mountain range and named at Cambridge University.
2 Off the coast of Cornwall, England, the marine biologist Paul Naylor has spotted the small species *Pagurus forbesii* using sponges instead of shells, and underwater photographer Jackie Hildering has photographed many examples.

The Lessons of Stone: Lake Van, Turkey

1 'Primordial oceans had oxygen 250 million years before the atmosphere', University of Minnesota bulletin on research by Mojtaba Fakhraee and Sergei Katsev, ScienceDaily.com, 25 January 2018.
2 It is even possible that Ulysses/Ulixes, the Latin name of Homer's hero Odysseus, was derived from the Hittite name Ullu(ya)š, which also means 'the far distant': we can see how this nickname ties in with the tale of Odysseus's *nostos*, or long sea-voyage back home. Odysseus itself might originate in a pronunciation of the logogram (or pictorial symbol) for Ut-napishtim, which would have been 'spelled' UD.ZI, but this may be unlikely.

3 Early mentions of Atrahasis are older still. A text now referred to as *The Instructions of Shuruppak* describes the handing down of wisdom to Ziusudra from his father Shuruppak, and has been dated to the early third millennium BC, or almost 5,000 years ago.

My God! Seals!: Farne Islands, England

1 As quoted in Philippe Diolé, *The Undersea Adventure*, Readers Union with Sidgwick & Jackson, London, 1954, p. 183.

Transfer to Reality: Hansankeyf, Turkey; Lion City, China

1 'With a Long Breath' by Zaradachet Hajo, published alongside the essay 'Portal to Antiquity: Hasankeyf, Turkey', by Joanne Leedom-Ackerman, in *World Literature Today* (2009), vol. 83, no. 4, p. 59.
2 Some scholars argue that the Critias of Plato's dialogues was not Critias the student of Socrates, but his grandfather, the elder Critias.
3 In his *Life of Solon* Plutarch tells us that in Heliopolis Solon spoke with a priest called Psenophis, and that in Sais he met the elderly priest who Plato had mentioned. He takes it for granted that the story was historical. Plutarch was – perhaps unwittingly – putting false flesh on the bones that Plato had provided. Plutarch's life coincided with the Roman Empire reaching its greatest extent, and his writings helped guarantee that the Atlantis story would remain imprinted on the Western mind.

The Floating Rocks: Santorini, Greece

1 Georges E. Vougioukalakis, *The Minoan Eruption of the Thera Volcano and the Aegean World*, Society for the Promotion of Studies on Prehistoric Thera, Athens, 2006, p. 5.
2 The exact date of the eruption is a subject of active debate. Radio carbon dating methods, ice-core dating and the dendrochronological

method (which refers to irregular growth rings in trees caused by volcanic winters) appear to have converged on a date near the end of the seventeenth century BC but this remains uncertain.
3 Tom Holland, speaking on *The Rest Is History* podcast, episode 315, 'Atlantis: Legacy of the Lost Empire', with Dominic Sandbrook, originally released 23 March 2023.

The Three Nails: Cabo de Palos, Spain

1 In *Letters to a Young Poet*, BBC Radio 3, originally broadcast September 2014.
2 Bottlenose dolphins belong to the *Tursiops* genus. *Tursiops* fossils first appear in the fossil record about 5 million years ago.
3 Stephanie King and Vincent Janik 'Bottlenose dolphins can use learned vocal labels to address each other', *Proceedings of the National Academy of Sciences of the United States of America*, 22 July 2013.

The Ruins: Yonaguni, Japan

1 The irony of von Däniken's attempts to make ancient human history sound more remarkable by attributing its most spectacular achievements to extraterrestrials is self-evident.
2 Graham Hancock, *Underworld: Flooded Kingdoms of the Ice Age*, Penguin Books, London, 2003, p. 647.
3 This is partly due to Schoch's theories about the Sphinx being thousands of years older than most antiquarians estimate, based on the idea that the horizontal striations on its body represent water erosion marks. The Sphinx is generally accepted to have been built during the reign of the pharaoh Khafre during the Old Kingdom Period, *c.*2570 BC. See Gabriel Moshenska, 'Alternative archaeologies', in *Key Concepts in Public Archaeology*, edited by Gabriel Moshenska, University College London Press, London, 2017, p. 133. Moshenska refers to Graham Hancock as 'the journalist Graham Hancock'.

11:43 a.m., 7 June 1692: Port Royal, Jamaica

1 The use of this language is important, as the word 'escaped', often used to describe slaves who had freed themselves from bondage, implies an equivalence with prisoners, rather than people whose only crime was being alive.
2 Jamaica Kincaid, *A Small Place*, Farrar, Straus & Giroux, New York, 1988, p. 14.
3 Patrick Whittle/Associated Press, 'Intense global overfishing is taking seafood dishes like conch in the Bahamas off the table', USA Today, 6 April 2023.

The Turn of Poseidon: Achaea, Greece

1 Homer, *Iliad*, 2.575, 8.203 and 20.404.
2 Speaking on *Helike: The Real Atlantis*, BBC *Horizon* documentary, originally broadcast 10 January 2002.
3 Pausanias, *Description of Greece*, Volume I: Books 1–2, 7.24.5–7.24.6, translated by W. H. S. Jones. Loeb Classical Library 93, Harvard University Press, Cambridge, MA, 1918.
4 Timaeus: 25a–25b, in *Plato in Twelve Volumes: Volume 9*, translated by W. R. M. Lamb. Harvard University Press, Cambridge, MA; William Heinemann Ltd, London, 1925.
5 D. Katsonopoulou, 'Helike and Her Territory in Historical Times', *Pallas* (2002), no. 58, p. 181.

He Walketh Through Dry Places: Pavlopetri, Greece

1 Speaking on BBC Radio 4, *In Our Time*: *Plato's Atlantis*, 22 September 2022.
2 An *ephor*, in modern Greek, means a government official, often one who is responsible for public works. The word has survived from archaic Greek times, where it meant one of the five chief magistrates of ancient

Sparta, who were so powerful and esteemed that they were the only citizens not required to kneel before the city's two kings.

Future Atlantises: The Coasts of a Rising Ocean

1 NOAA/climate.gov, *Climate Change: Global Sea Level*, Rebecca Lindsey, reviewed by R. Lumpkin, G. Johnson, P. Thompson, W. Sweet, originally published 19 April 2022.
2 Ibid.
3 Ibid.
4 Aysha Imtiaz, 'The unlikely protector against Bangladesh's rising seas', BBC, 1 September 2021.
5 Joe Phelan, 'What countries and cities will disappear due to rising sea levels?', LiveScience.com, 27 March 2022.
6 Ibid.
7 Mayuri Mei Lin and Rafki Hidayat, 'Jakarta, the fastest-sinking city in the world', BBC Indonesian, 13 August 2018.
8 Shamsuddin Illius, 'Stability eludes climate refugees in Bangladesh's sinking cities', PreventionWeb.net, 25 January 2023.
9 Quoted in Joe Phelan, 'What countries and cities will disappear due to rising sea levels?', LiveScience.com, 27 March 2022.
10 These can include coir or coconut-fibre logs or oyster-shell sandbags.
11 M. S. N. Chowdhury, B. Walles, S. M. Sharifuzzaman, M. Shahadat Hossain, T. Ysebaert, A. C. Smaal, 'Oyster breakwater reefs promote adjacent mudflat stability and salt marsh growth in a monsoon dominated subtropical coast', National Library of Medicine/National Center for Biotechnological Information, 12 June 2019.
12 Ibid.
13 Ibid.
14 Aysha Imtiaz, 'The unlikely protector against Bangladesh's rising seas', BBC, 1 September 2021.

Epilogue: Lokroi, Greece

1 Thucydides, *The Peloponnesian War*, 3.89.3, J. M. Dent, London; E. P. Dutton, New York, 1910.
2 Iain Stewart, 'Echoes of Atlantis', *Guardian*, 26 October 2000.

Index

Abraham 69
Accompang (Maroon leader) 197
Acanthurus coeruleus (blue tang) 200
Achaean League 216
Achilles 164
Acropolis of Athens 44, 88-9, 92
Adam 67
Aestivation 192
Aetobatus narinari (spotted eagle ray) 202
Agamemnon 216
Aigio, Greece 218, 221
Akrotiri (Santorini), ruins of 143–5
alcohol 195
Alexandria, Egypt 67
Aliger gigas (queen conch) 210–11
Alighieri, Dante 68–9
alveoli 107
anemones 73, 102, 132
Animal Welfare (Sentience) Bill, UK 151
Anolis aquaticus 127
anthropomorphism 234
Antikythera Mechanism 170–2, 212
Antonia Minor 162
Arabs 136
Aral Sea 135
Arganthonios of Tartessos 55
Aristotle 45
Arslantepe, Turkey 50
Art of the Aqualung, The (Gruss) 129
Asterias rubens (common sea-star) 97
astronomical clocks 171
Atalanta, Greek heroine 244–7
Atalantonisi, Greece 245–6

Atlantic Ocean
  and Celtic mythology 68
  and conditions in English Channel 95
  and continental drift theory 65, 72
  limits of 47
  location of Atlantis in 4, 43, 48, 90, 142
  name of 5, 42
  in poetry of Blake 68
  theory of lost lands in 87
  sea rises in 239
  seabed of 92, 153
Atlantis
  aesthetics of 88
  army of 44
  artistic imaginings of 67
  colonialism and 47–9, 67
  conflict with Athens 5
  debates about reality of 45
  expeditions in search of 92–3, 216
  geography of 4–5
  literalism and 215
  modern connotations of 12, 45
  navy of 44
  Nazis and 50–1
  New World as 48
  origins in Plato 41–5
  resurgence of interest in 46
  supposed earlier mentions of 42
  theory of Andalusia as 48
  theory of Crete as 89
  theory of Santorini as 50, 141–2, 146
  theory of Sweden as 49

Atlantis Massif 153
Atlantis the Palm Hotel, Dubai 139
Athena 44
Athens, Greece
   aesthetics of Atlantis and 89
   conflict with Atlantis 138
   'golden age' of 142
Atlas, son of Poseidon, King of Atlantis 4, 42
Atlas the Titan 4, 42
atlases 65–6
Atrahasis epic 110
Aulostomus maculatus (West Atlantic trumpetfish) 198
Auschwitz 41
Austrothelphusa transversa (the Australian inland crab) 192

Baal Hammon 181
Bahamas, the 210
Baia, Italy 155–62 *passim*
Balistes vetula (queen triggerfish or old wife) 200
Bangkok, Thailand 241
Bartoli, Giuseppe 49
'bends, the' *see* decompression sickness
Bermuda 199
Bible, the 49, 67, 109, 231
Blake, William 68–9
Blavatsky, Helena 86
Blondel, Paul 196
Blue Mountains, Jamaica 195
Boston University 190
bradyseism 157
Brendan the Navigator, St 68
British Broadcasting Corporation (BBC) 243
British Empire 51, 197–8
British Sub Aqua Club (BSAC) 22
Brontë, Emily 138
bronze 50, 55, 171

Bronze Age 55, 142–3, 146, 216, 225–6, 230–1
Brutus of Troy 69
bubbles 16, 31, 36–8, 122–3, 160, 170, 182, 202, 215, 234
   as giveaway of diver's position 85, 128
   as repellant to marine animals 117
Buccinum undatum (common whelk) 102
buoyancy control devices (BCDs) 31–2, 35–7, 56–7, 79, 81–5 *passim*, 96, 98, 122, 123, 148, 169, 201
Byron, Lord George Gordon 219
Byzantines 46, 136, 218

Cabo de Palos, Spain 181
Cadiz, Spain 4, 54
Caesar, Julius 69
caffeine 117
Cameron, James 93
Campania, Italy 155
Campi Flegrei, Italy 156
Canada 109
Cantabrigiaster fezouatensis 98
Capra hircus cretica (kri-kri ibex or Cretan goat) 244
Caranx hippos (crevalle jack) 202
carbon dioxide 128
carbon monoxide 62
Carboniferous period 66
Cardiff University 171
Cartageña, Spain 54, 180, 186
Carthaginians 54, 181
Cave of Heracles, Greece 221–2
Cayman Trader (wreck of) 200–3
Cerberus 155
Çeşme Bay, Turkey 141
Chaetodon sedentarius (reef butterflyfish) 200
Chattogram, Bangladesh 241

Chesapeake Bay Environmental Centre 242
Chiang Rai cave rescue 21–2
Chowdhury, Mohammed Shah Nawaz 243
Chromis chromis (Mediterranean chromis, or damselfish) 183
cist graves 230–2
City of Waterford, wreck of 165–6
Claudius, Roman Emperor 162
Cleito of Atlantis 4, 44, 50, 247
cocaine 209
Coleridge, Samuel Taylor 138
Colpophyllia natans (boulder brain coral) 200
compass (drawing tool) 221
compass (navigational) 58, 83–4, 170
Conger conger (European conger eel) 174, 184
Corinth, Gulf of 214–27 *passim*
Cortés, Hernán 47
Cousteau, Jacques-Yves 16, 104, 128–130, 148, 216–223 *passim*
COVID-19 95
Crantor of Soli 45
Critias the Elder 137
Critias the Younger 137
Cudjoe (Maroon leader) 197
cuneiform script 110
cylinders (scuba) 32–3, 35, 56, 77, 112
Cyphoma gibbosum (flamingo's tongue sea snail) 201
Cyprinus carpio (common or Eurasian carp) 7
Cyprus 92

Dalley, Stephanie 110
Dawlish Warren, England 238
de Neuville, Alphonse-Marie-Adolphe 182

decompression sickness ('the bends') 33–4, 76, 111
decompression stops 77, 96, 121, 147
deoxyribonucleic acid (DNA) 171
depth
  of deep ocean 59–60
  etymology of 130
  exceeding limits of 148
  limits for different divers 121
  technical diving and 76
Deutsches Ahnenerbe 51
Dhaka, Bangladesh 240–1
Diadema antillarum (black sea urchin) 206–7
Dicentrarchus labrax (European seabass) 102
Dickens, Charles 9
Die Hard (1988) 82
dimethyl sulphide 116
Diodon hystrix (spotted porcupinefish) 202
Diogenes pugilator (Roux's hermit crab) 102
Diogenes the Cynic 102
Diolé, Philippe 16, 129–30, 147
dive computers 32, 77, 84, 111, 123–4, 126, 168, 170, 174–5
dive gear, allure of 18
dive masks
  and claustrophobia 39–40, 179
  clearing water from 34–5, 58, 79
  older designs of 128
  preventing fogging of 97
Divine Comedy, The (Dante) 68–9
diving watches 171–2
Donadey, Claude 208
Donnelly, Ignatius 86–7
Dropides of Athens 137
druids 69
drysuits 18, 78–83 *passim*, 117, 121–3, 125, 201
Dubai, United Arab Emirates 139

East Carolina University 242
Echidna catenata (chain moray eel) 211
Eden, garden of 67
Edgerton, Harold 'Doc' 216–17
Egypt
   antiquity of 137
   flooding of 239
   Herodotus and 138
   Plato's use of 87
   priests of 5
   pyramids of 191
Elafonisos, Greece 229
Elizabeth I of England 9
Elizabeth II of England 187
English Channel 112, 166
Ephorus of Cyme 55
Equetus punctatus (spotted drum) 200
Erdoğan, Recep Tayyip 138
Esox lucius (northern pike) 7
Estartit, Catalunya 24
Eratosthenes 45, 217
Euboean Gulf, Greece 244
eustachian tubes 38

fear
   of diving 16, 21
   of drowning 38
   of entanglement 23, 94
   of fascists 41
   of humans 101–2
   overcoming fear 17, 23, 174, 236
   of panic 173
   of poor visibility 83
   of runaway ascent 84
   of running out of air 63
   of sea creatures 184
   of shipwrecks 173
   of the sea 3, 17, 114, 237
   of volcanoes 156
Feder, Kenneth 91

fins (scuba) 16–17, 36
flamenco 179
flatfish 25
Fleuss, Henry 128
Flying Dutchman 62
Fort Charles, Jamaica 197-8
Fort James, Jamaica (ruins of) 207–9
Frankfurt 71
'frogmen' 128
Fucus spiralis (spiral wrack) 11

Gagnan, Émile 129
Galathea squamifera (common squat lobster) 12, 132
gas narcosis 147–8
geosynclinal theory 71
Gerartsen van Gorp, Jan 47
German Geological Association 71
ghost ships 62
giants 69
Gibraltar, Strait of 90
Gilgamesh 135–6
Gill, Christopher 229
gin 209
Ginglymostoma cirratum (nurse shark) 201–3
Gittman, Rachel 242
Glaucus atlanticus (blue sea dragon, blue angel or sea swallow) 61
Godfrey-Smith, Peter 149
Gondwanaland 66
Gorgonia ventalina (common sea fan) 200
gneiss 109
Grapes, The (public house) 9
Graves, Robert 89–90
Gruss, Robert 129
Guadalquivir river, Spain 54
Gulf of Corinth 214–27 *passim*
Gunan, Saman 22

Hades 164
Haemulon flavolineatum (French grunt) 211
Halichoerus grypus (grey seal) 131–3
Halisarca caerulea (thin encrusting sponge) 208–9
Hancock, Graham 92, 189–90
hand signals (scuba) 58, 160
Harding, Anthony 229
Hardy, Thomas 210
Harvard University 7
Hasankeyf, Turkey 136–9 *passim*
Hellenistic period 224–5
Helike, Greece 215–27 *passim*, 229
'heliox' 76
Hephaestus 44, 141
Herculaneum 156
Herodotus 42, 44, 55, 138, 229
Hesiod 42
Hetzler, Florence 209
Himanthalia elongata (thongweed) 100
Himmler, Heinrich 51
Hipparchus of Rhodes 171
History of the Peloponnesian War (Thucydides) 245
Holland, Tom 45
Homarus gammarus (European or common lobster) 122, 175
Homer 216, 224
hoplites 50
hunting 24–5, 98, 175, 190
Hyas araneus (great spider crab) 12, 112, 132
Hydractinia echinata (hermit crab hydroid) 103
hydrodynamics 132, 235
hydrogen 62
hydrogen sulphide 106

hydrothermal vents 153
hyperdiffusionism 70–1

Iliad, The (Homer) 41, 216
Ilisu Dam, Turkey 136
Illius, Shamsuddin 241
Indicopleustès, Cosmas 67
Irish Sea 166
Islamic world (preservation of Western texts in) 87

Jacobovici, Simcha 93
Jakarta, Indonesia 240
Jamaica National Heritage Trust 199, 205
James, M. R. 138
Janik, Vincent 184
Jebel Musa, Morocco 90
jiandu chronicles 140
Jolly Roger 198
Jonah 183

Kaplan, Robert 91
Katrina, Hurricane 239
Katsonopoulou, Dora 219–26 *passim*, 246
kelp 131
Kereinitis River, Greece 216
Kiel University 189
'kill devil rum' 209
Kimura, Masaaki 189
Kincaid, Jamaica 199
King, Stephanie 184
Kingston, Sussex, England, old chapel of 2
Kingston, Jamaica 194–5
Kiribati 240
knives (divers') 18, 25, 31
knot rings 1
Koprinka reservoir, Bulgaria 241
Kos, Greece 141

Kronfield, Paul 217
Kurdistan 136
Kutubdia Island, Bangladesh 243

lagoons 219–20
Lagos, Nigeria 241
landfill 62
Lang, Fritz 12
Latin 54, 155, 157, 166
Laurasia 66
lead (metal)
    boots of 88
    price of 126
    weights of 35, 37, 55, 63, 81, 125
Lego 203
Lemuria 86
Lewis, C. S. 152–3
Lime Cay, Jamaica 201
Limehouse Reach, England 9
Link, Edwin 196
Lipohrys pholis (common blenny, or shanny) 12, 102
liquefaction 220–1
London School of Economics 151
López de Gómara, Francisco, 47–8

Mac Lir, Manannán 68
Maia, daughter of Atlas 42
Maja squinado (European spider crab) 12, 112, 132
Maldives, The 240
Malea peninsula, Greece 228
mangroves 195, 201, 242–3
Mar Menor, Spain 180
Marinatos, Spyridon 142–5
Marx, Robert 199, 209
Mary Celeste, the 62
Masselink, Gerd 242
Matthew's gospel 231
Mazzoldi, Angelo 70

Medes 136
Medway, River 23
Mellisuga minima (vervain hummingbird) 195
mermaids 132
Mesolithic humans 190
Mesopotamia 109
Mesozoic era 66
Methana, Greece 141
microbialites 107–9
Millepora (fire corals) 200
Mnemosyne 191
Mongols 136
mosaics 160–1
Muraena helena (Mediterranean moray eel) 232
mythology
    Atlantis as 92
    Greek 16, 216, 244–7
    Iberian 54
    Irish 26, 68
    Norse 26
    Roman 156

nails 186–7
names
    of bottlenose dolphins 184
    of fish 184
Nanny of the Maroons 197
Naples, Italy 154–5
Nautilidae 193
Nazis 49–51, 166
Necor puber (velvet swimming crab) 101, 125
neoprene 26, 76, 117
Netherlands, The 242
New York City 241
Nikolaiika, Greece 216
Noah 67, 69
Noffke, Norah 106
Northumberland, England 115

nudibranchs 60–1
Númenor, fictional island of 138

Oblada melanura (saddled sea bream) 159, 161, 185, 234
Ocean View dive club, Sussex, England 94, 166
Octopus vulgaris (common octopus) 60, 131, 149–50
Odysseus 164
Old Naval Hospital, Jamaica 206
Old Norse 119
Olde Ship Inn, the 116
Ordovician period 207
Orihuela Costa, Spain 30, 179
ostracods (seed shrimp) 222–3
Otranto, Strait of 90
Ottomans 136
oxygen
  absence from ancient atmosphere 106–7
  and early rebreathers 128
  sea's generation of 107
  solubility of 31
  toxicity at depth 76, 128
oxygen toxicity 76
oysters 242–3
Öztürk, Ali Ihsan 108

Pacific Ocean 239–40
Pagurus bernhardus (common hermit crab) 100–1
Palaemon serratus (common prawn) 12
Pangaea 66
Paterson, Don 181
Pausanias 217
Pavlopetri, Greece 228–236 *passim*
Pegwell Bay, Kent 69
Pelecanus occidentalis (brown pelican) 195–6
Pellicer de Ossau y Tovar, José 48

penises 222
permeable pavements 242
Phocis, Greece 224
Phoenicians 54
Pholidichthys leucotaenia (engineer goby) 73–4
pia mater 9
pirates 195, 197, 209
Pilbara, the, Australia 106–7
Pillars of Heracles (Hercules) 4–5, 43, 90–1, 221, 246
plasma arc gasification (PAG) 62
plastic 61–2, 150, 185, 211
plate tectonics 72
Plato
  as dramatist 42
  as humorist 247
  ignorance of prior history 228
  loss of works in the West 46
  Timaeus and *Critias* of 41–6 *passim*, 216
  transmission of Atlantis story to 6, 137
Pliny the Elder 181
Pompeii 156
poppies 111
Port Royal, Jamaica 195–213 *passim*
Poseidon 4–5, 44, 48, 50, 69, 214–7, 223
Posidonia seagrass 148, 231
Posidonius 45
potash 128
Pounta, Greece 229
Procellaria glacialis (northern fulmar) 119–20
Proclus 45
Professional Association of Diving Instructors (PADI) 16, 23
prostitution 195
pseudo-archaeology 142

Qiandao Lake, China 138–9
Quao (Maroon leader) 197

Rán, Norse goddess of the sea 2–3, 114
rebreathers 117, 126, 128, 133, 175
Red Book of Hergest 138
regulators (scuba)
    dangers of 25, 34–5
    recovering when lost 38
    sensation of breathing from 37
    types of 32–3
Riou, Édouard 182
RMS Titanic 241
rockpools 11–12
Roerich, Nicholas 189
Rome (ancient)
    founding of 42
    fall of 46
    rivalry with Carthage 54
    role in Hellenisation 70
    survival of Eastern half of empire 46
Romans 51, 136, 158, 181, 224
Romany Gypsies
    fear of the sea 17
    and belongings of the dead 104
    origin myths of 178
Royal National Lifeboat Institution 77
Rudbeck, Olof 49

safety stops 96, 121, 124, 148, 175–6
Sahara, dust of 179
Sais, Egypt 5, 46, 88, 137
salt 61, 64, 76, 163, 191
salt marshes 242
Salvation Army 15
Santorini, Greece
    identification of Atlantis with 142
    volcanic explosion on 141
Sarmast, Robert, 92–3
Sarmiento de Gamboa, Pedro 47
saturation diving 105
Saturn (Roman god) 181
Schliemann, Heinrich 52, 69
Schoch, Robert 190

Schulten, Adolf 55
Sclater, Philip 86
Scomber scombrus (Atlantic mackerel) 25–6
Scylla and Charybdis 90
Seahouses, England 115–6
sea cucumbers 207, 210
'sea-peoples' 89
sea level rises 237–43 *passim*
sea walls 118, 238, 242–3
Selinountas river, Greece 223
selkies 26, 33, 36, 132
Seuthes III of Thrace 241
Shadwell Basin, England 7–8
Shakespeare, William 61
sharks 26, 31–2, 188, 201–3
sheep 82
Shelley, Mary Wollstonecraft, *née* Godwin 138
Shicheng (Lion City), China 138–9
Ships
    of Agamemnon 216, 224
    of Atlantean navy 44
    and damage to seabed 235
    and hubris of explorers 227
    plunder of 195
    of 'unlimited range' 62
shipwrecks
    of the Aral Sea 136
    as corpses 169
    of the English Channel 165
    of the Farne Islands 132
    love of some divers for 175
    swimming through 185
    as tombs 167
Sicily, Strait of 90
Silent World, The (film: 1966) 16
Silent World, The (Cousteau) 128
Smaal, Aad 243
snorkelling 232–3
Socrates 137

Solon of Athens 137–8
Soter, Steven 221
Southern Ocean 239
Sparta 51
Sparus aurata (gilt-head or silver seabream) 159
Sprague de Camp, Lyon 246
Sprawson, Charles 219
SS Fortuna, wreck of 165–77 *passim*, 187
St Kilda 119
starfish 98
Stewart, Iain 245
Stingray (TV show) 11
stones
    of Atlantis 91, 142
    birth of 153
    as indicator of human activity 189–90
    octopus' use of 149
    paying attention to 112–3
    in rockpools 11
Strabo 45
subarachnoid haemorrhage (stroke) 9
Sumerian King List 229
surface marker buoys (SMBs) 120–1, 124
Sybaris, colony of Helike 216
Swierk, Lindsey 127
syngas 62

tabulate corals 191
Tailliez, Philippe 127, 130
Taíno people 197
Tanit 181
Tarshish 55
Tartessos 55
tears 163
tels 45
Tempest, The (Shakespeare) 61
Tenochtitlan, Mexico 48
Tesson, Sylvain 183

The Origin of Continents and Oceans (Wegener) 71–2
Thucydides 229, 245
Tibet, architecture of 189
Tigris (river) 136
Tilev, Zheko 241
tin 55
Tír na nÓg ('Land of Youth') 68
Tolkien, J.R.R. 46, 138
Torrevieja, Spain 29
treasure hunters 2, 70, 199
Triassic period 207
trim (scuba term) 58–9
triremes 50
Troy 41, 52, 69, 224
tsunami 141, 195, 221, 245
Tursiops truncatus (common bottlenose dolphin) 184
Twenty Thousand Leagues Under the Sea (Verne) 88
Tyrrhenian Sea 163

USS Gerald R. Ford 62
UNESCO 199
Union of Concerned Scientists, The 240
University of Chittagong 243
University of Marburg 65
University of Plymouth 242
University of St Andrews 184
University of Strasbourg 51
Upnor, England 23
Ut-napishtim 109–10

Vacelet, Jean 208
Van (lake), Turkey 107–9
Vatika Bay, Greece 235
Verne, Jules 88
Vesuvius, Mount 156
Vidal-Naquet, Pierre 41, 49–51 *passim*, 67, 142

volcanoes 141, 144, 156
von Däniken, Erich 189
Vougioukalakis, Georges 141
Vulcan 156

Wageningen University 243
Walters, Selvenious 205
Wegener, Alfred 65–72 *passim*
Weil's disease 7
wetsuits 18, 22, 26, 30–1, 33, 57, 63–4, 80, 95, 145
Whitehead, A. N. 42
Widewater Lagoon, England 95, 103
Wirth, Herman 50–1
Wrackers 199

Wraysbury, England 19, 75, 78, 82–3, 99
Woods Hole Oceanographic Institution of the United States 92

x-rays 196
Xerxes I (called the Great) 44
Xestospongia muta (giant barrel sponge) 200
Xin'an River, China 139

Yonaguni, Japan 188–91

ziggurats 108

About the author

Damian Le Bas is a writer, film-maker and visual artist. His first book *The Stopping Places* won the Somerset Maugham Award, a Royal Society of Literature Jerwood Award, and was shortlisted for the Stanford Dolman Travel Book of the Year. Le Bas is widely published as a journalist and poet and has taught for the Arvon foundation. A recipient of a Society of Authors Travelling Scholarship, he is a member of the European Film Academy, holds a First Class degree in Theology from the University of Oxford and was awarded an honorary Master of Education by the University of Chichester. Besides getting in the sea he loves music, walking and spending time in the woods and hills with his family and friends. *The Drowned Places* is his second book.